AF769931

THE INVISIBLE BOMB

The Nuclear Arms Race in the Middle East

FRANK BARNABY

I.B.TAURIS & CO LTD
Publishers
London

Published by
I.B.Tauris & Co Ltd
110 Gloucester Avenue
Primrose Hill
London NW1 8JA

Copyright © 1989 by Frank Barnaby.

All rights reserved. Except for brief quotations in a review, this book, or any part thereof, must not be reproduced in any form without written permission from the publisher.

British Library Cataloguing in Publication Data

Barnaby, Frank
The invisible bomb: nuclear weapons and the Middle East.
1. Middle East. Nuclear weapons. Policies of governments
I. Title
355'0335'56

ISBN 1-85043-078-0

Typeset by Columns of Reading.
Printed and bound in Great Britain by
Redwood Burn Limited, Trowbridge, Wiltshire.

Contents

Combat radii of various aircraft from air bases in Israel in kilometres

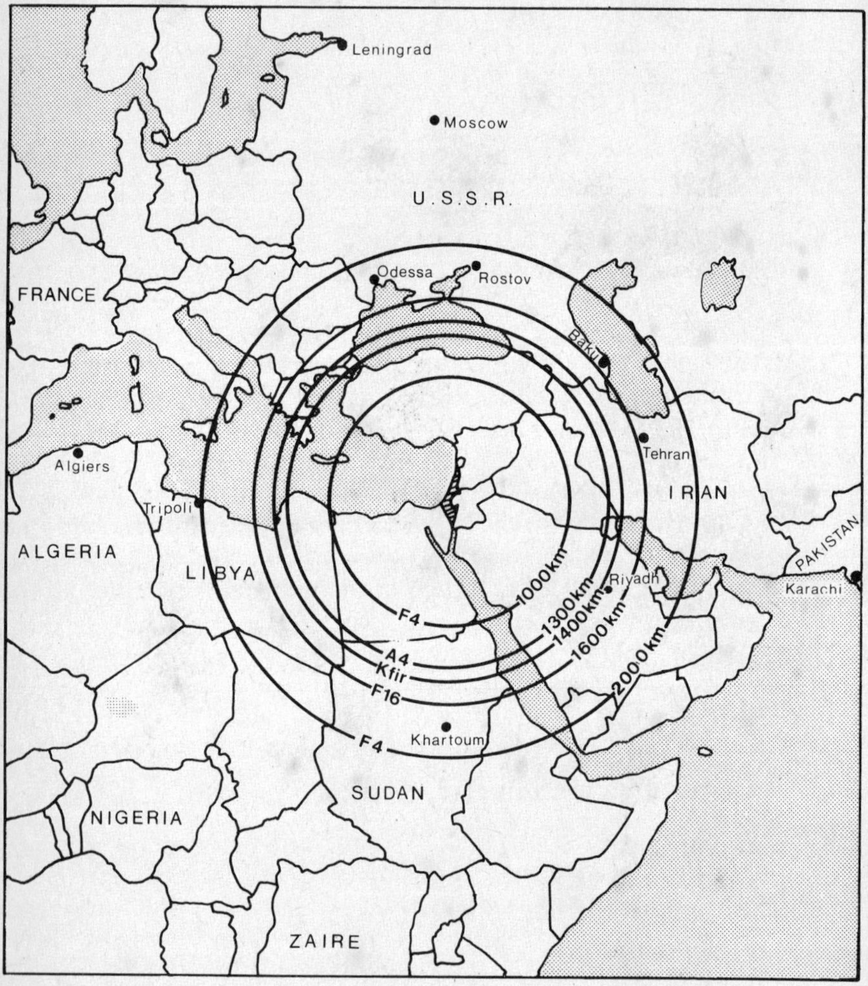

Note: The ranges for one-way missions and less than maximum weapon loads are more than twice the combat radii. The outer circle is the one-way range of the F-4.

Preface

Any person with even a vague knowledge of the history of the Jews will be aware of the dreadful persecution this remarkable race has suffered over the past 2000 years. And frequent events bring home to us that anti-Semitism is still very much alive. Given these experiences, the Jews must assume that fascist, or other anti-Semitic totalitarian, regimes will yet again find that it suits their ends to persecute them. It would, then, be rash for them to believe that there will not be another Holocaust. And it would be even more rash to believe that others will be more willing to come to their aid next time than they were when Hitler was exterminating 6 million Jews. The only safe haven today's Jews have is the State of Israel.

In today's Middle East, Israel – with a population of 4.3 million – feels secure only if it is armed with the most powerful weapons its scientists can produce. We may know that the United States will not allow Israel – its most reliable ally, and the only real democracy, in the world's most important strategic region – to be destroyed. But, given the history of the Jews, few Israelis rely on so-called 'guarantees'. And who can blame them?

Because they are unwilling to rely on others, Israelis want to be self-sufficient in weapons production – both in conventional and nuclear weapons. This desire is enhanced by the willingness of arms-exporting countries to sell the most modern conventional weapons to Israel's adversaries – such as the huge deals made recently by the British to sell advanced combat aircraft to Saudi Arabia.

Israel's attitude to nuclear weapons is influenced by the attitude of NATO and Warsaw Pact countries to these weapons. NATO leaders, for example, frequently claim that nuclear weapons are needed to keep the peace in Europe. If this is true for Europe, it is equally true for Israel. And Israel is bound to be influenced in its nuclear decisions by the knowledge that Arab countries – such as Iraq and Syria – are producing chemical weapons of mass destruction and that Iraq has shown itself prepared to use them in war.

However understandable it may be that Israel has produced nuclear weapons, the fact remains that the presence of nuclear weapons in the Middle East is a grave threat to world security. The most likely way in which a nuclear world war will occur is, after all, through the escalation of a local nuclear war in the Middle East.

The proliferation of nuclear weapons in the Middle East is clearly a crucially important issue, perhaps the most important current international issue. I was, therefore, most interested in what Mordechai Vanunu had to say about Israel's nuclear capabilities during the time I spent with him after the London *Sunday Times* brought him from Sydney to London. Because of lack of space, *The Sunday Times* could only summarize Vanunu's story. It is important that his information is made widely available, and this is the purpose of the first part of this book. I can only report what Vanunu said; I have, of course, no way of checking his information.

I am extremely grateful to *The Sunday Times* for the opportunity to talk to Vanunu, and particularly to Peter Hounam, of that paper, for the help he has given me in collecting the information for this book. I would also like to thank Anna Enayat and Margaret Cornell for their excellent editorial work.

1
Introduction

In March 1988, Mordechai Vanunu, a 33-year-old Moroccan-born Israeli, was found guilty by the Jerusalem District Court of treason, aggravated espionage and the collection of secret information with intent to harm Israel's security. Vanunu was sentenced to 18 years' imprisonment.

He was accused of these grave crimes because he gave the London-based *Sunday Times* top secret information about Israel's production of materials for use in nuclear weapons – information collected when he worked as a technician from 2 November 1976 to 27 October 1985 at Israel's Dimona Nuclear Centre in the Negev desert. Some of this information was published on 5 October 1986 in an article called 'The Sunday Times Reveals: The Secrets of Israel's Nuclear Arsenal'. This book describes in some detail the information about Dimona provided by Vanunu.

The Israeli authorities admit that Vanunu was employed at the Nuclear Research Centre in the Negev near Dimona as 'a technician and operator' from 2 November 1976 until 27 October 1985. When he had finished his course of instruction at the start of his work at the Centre, he signed an affidavit to maintain secrecy. He was also interviewed by the Centre's security officers and told about his commitments to maintain secrecy. According to the prosecution at his trial, he was 'also instructed from time to time about the need to maintain secrecy and signed a number of affidavits and commitments to that effect'.

The prosecution alleged that while he worked at the Dimona Centre, mainly from the beginning of 1985, Vanunu collected, prepared, recorded and held in his possession secret information – all without being authorized to do so and with intent to impair the security of the state'. It also admitted that he visited, without authorization, various top secret areas in the Dimona Centre and 'photographed various installations and objects and also copied details and diagrams from professional booklets, the contents and security grade of which are secret'. The information collected included information on secret technical developments, operating procedures and production processes at the Centre, and the code names and terminology of various secret developments.

On 19 January 1986, Vanunu left Israel, taking with him 57 photographs and the handwritten notes which he had made at Dimona. In May 1986, he arrived in Sydney, Australia, where he met Peter Hounam, a *Sunday Times* journalist, and told him about his work at Dimona and the information, including the photographs, he had collected there.

Vanunu was brought from Sydney to London where he was kept in secret locations. By the time the story appeared in *The Sunday Times* he had disappeared from London. On 30 September 1986 he was lured to Rome and then, it is believed, kidnapped by agents of Mossad, the Israeli secret service, and taken by ship to Israel.

At the request of *The Sunday Times* I spent two days in September 1986 cross-examining Vanunu. It was clear to me that he had, as he claimed, been a technician working on several processes in Dimona. He was very straightforward about what he did and did not know and made no attempt to comment on matters outside his experience. I found his story totally convincing, and the photographs he brought with him considerably increased his credibility. Other nuclear physicists – particularly the eminent American nuclear-weapon designer, Theodore Taylor – who checked Vanunu's story also found it credible. After Vanunu's evidence, backed up by photographs, there can no longer be any reasonable doubt about Israel's nuclear capability.

In the last few months of 1987 Vanunu was tried in the Jerusalem District Court – the trial was held *in camera* – and accused of the following violations of Israeli law:–

i) treason – assistance to an enemy in war – a violation of Section 99 of the 1977 Penal Law;
ii) aggravated espionage – delivery of secret information with intent to impair the security of the state – a violation of Section 113(b) of the 1977 Penal Law; and
iii) collection of secret information with intent to impair the security of the state – a violation of Section 113(c) of the Penal Law.

The form the charges took can be seen as an official admission that the information given by Vanunu – or, at least, much of it – is true. Yet, after *The Sunday Times* article appeared, Shimon Peres, the then Prime Minister of Israel, repeated the standard Israeli formula that 'Israel would not become the first country to introduce nuclear weapons into the Middle East' – a statement which seems to be a direct denial that Israel has nuclear weapons. (Incidentally Peres should know many of the details of Israel's nuclear programme. As Director-General of the Ministry of Defence in the early 1960s he is generally regarded, together with Ben Gurion, as the architect of Israel's nuclear-weapon programme.)

Nevertheless Vanunu was not charged with telling lies about his country's nuclear activities. He was charged with treason and espionage. This must mean that his statement that Israel has produced the nuclear components for a relatively large number of nuclear weapons is true. The Israelis may not have actually assembled their weapons, but they could do so in a very short time.

If the Peres statement means that Israeli keeps the fissile components (i.e., the plutonium) of its nuclear weapons outside the rest of the assemblies, and, therefore, does not technically have nuclear weapons, this is a quibble; it would take a very short time to screw the fissile components into the weapons. But if, as is likely, this quibble is the justification for Peres's statement he can at least argue that it can be supported by international law. So far as the Nuclear Non-Proliferation Treaty, for example, is concerned, 'a nuclear weapon' only becomes a nuclear weapon when the fissile material is inserted into the device. If the fissile component is kept just outside the weapon, then, according to the treaty, it is not 'a nuclear weapon' even though it is the standard practice of the nuclear-weapon powers to keep nuclear material separately in order to make the nuclear weapons more difficult to steal). So far as Vanunu is concerned, of course, Israel has nuclear weapons if it has manufactured the components for them, even if the fissile material is stored separately from the non-nuclear part of the weapon.

The fact that the Israeli Government went to great lengths to get Vanunu back to Israel and to put him on trial for giving away secrets about Israel's nuclear-weapon programme can only mean that it does not expect anyone to believe its denial about its possession of nuclear weapons. The point of the denial is that other governments, particularly the US Government, are able to maintain the myth that Israel does not have nuclear weapons. This is important for Israel because the US Government cannot, under US law, continue to give economic aid to Israel if it acknowledges that Israel has nuclear weapons.

In this context, it is bound to occur to many people that Vanunu's revelations, coming at the particular time they did, may well be convenient for Israel. Far from impairing the security of the state, they may, in fact, actually improve it. As will be seen, the Israeli authorities may want the Arab states to know that they have a sophisticated nuclear force. At the same time they are unwilling, for the reasons given above, to make the announcement officially.

Such an interpretation would also explain the behaviour of the Israeli security services. That Vanunu was able to smuggle camera and films into Dimona – Israel's most secret military establishment – and photograph highly secret areas in the plant implies an almost unbelievable incompetence on the part of the security services. That this happened when Vanunu was known to be critical of some aspects of Israeli policy compounds the incompetence.

Vanunu was then allowed to leave Israel openly and to travel, with his films and written notes, to Australia. While there, he discussed his work at Dimona and made it known that he had films of the secret processes there.

There is evidence that an Australian citizen with connections with the Australian secret service heard about this and alerted the Australian authorities.

Mossad probably heard about Vanunu's activities in Australia, possibly via the Australian Secret Service and the US Central Intelligence Agency. And still it did nothing. In fact, it only acted after *The Sunday Times* informed the Israeli embassy in London that it intended to publish details of Israel's nuclear-weapon programme given to it by Vanunu.

History shows that Mossad is a highly competent secret service. The most credible explanation of its behaviour in the Vanunu case is that it found out what Vanunu was up to and decided to give him the chance to tell his story. His abduction from London via Rome – the sort of highly efficient operation one has come to expect from Mossad – and his trial in Jerusalem have added considerable authenticity to Vanunu's account. This is just what one would expect if the Israeli authorities wanted the world to take notice of Vanunu's disclosures. If they had done nothing, the *Sunday Times* article would have very soon been forgotten. Instead, it has received considerable international attention in all forms of the media.

I am not suggesting for one moment that Vanunu was a willing tool of Mossad. On the contrary, my conversations with him convinced me that he was not. But it is entirely possible that unwittingly he was allowed to serve a purpose – to tell the world about Israel's nuclear-weapon activities. The Israeli leaders cannot admit that Israel is carrying out these activities but they are quite happy that Vanunu has done so.

VANUNU'S DEFENCE

At his trial Vanunu explained to the court that he violated Israel's secrecy laws only because he believed that both the Israeli and the world public had the right to know about the information he handed over. In other words, his motives for talking were ideological. Israel's political leaders have, he says, consistently lied about the country's nuclear-weapon programme.

Vanunu's defence is based on the danger that a future war in the Middle East may begin with conventional weapons, escalate to a local nuclear war in which Israeli nuclear weapons, for example, are used, and then develop into an all-out nuclear war between the superpowers. A glimpse of this possibility can be had from events in the 1973 Yom Kippur war in which Israel is believed to have deployed nuclear weapons, and the Soviet Union is understood to have sent nuclear weapons to Egypt, and probably Syria, for SCUD surface-to-surface missiles, and both superpowers put their nuclear forces on alert.

The USA and the USSR are inevitably involved in wars in the Middle East because they supply the bulk of the weapons used in these conflicts. Modern warfare uses weapons, particularly missiles, at a great rate, as was shown dramatically in the 1973 war. Within a few days, both sides ran short of weapons and were saved from defeat only by deliveries of Soviet and US arms in massive airlifts. The arms supplier thus becomes the guarantor of his client's survival. Neither superpower will readily allow a client to be defeated in war; to do so would be to lose much of its credibility as an ally. If the USA, for example, stood by while Israel was destroyed by the Arab countries, it would no longer be a credible partner in NATO.

If the superpowers are committed to opposite sides in a Middle East war, the potential for escalation right up to a strategic nuclear war is obvious. In October 1973, for example, President Nixon put the US Strategic Air Command on high nuclear alert to deter the Soviets from sending troops to separate the Israeli and Egyptian forces, thus making it clear that, on this occasion, he was willing to risk escalation to an all-out nuclear war. Any use of Israeli nuclear weapons would, to say the least, considerably increase the risk of such escalation.

There is, then, an obvious link between Israel's nuclear forces and world security. And this is why Vanunu believes that he was justified in telling the world about the amount of nuclear-weapon material being produced at Dimona. He maintains that the world public should know that Israel is producing material for a much larger nuclear force than is needed as a last-ditch deterrent to prevent the Israelis from being overwhelmed by Arab military forces and pushed into the sea.

Do Israel's nuclear activities, he asks, mean that its nuclear policy is changing from one of deterrence to the much more dangerous one of nuclear war-fighting? Vanunu argues that such crucial questions should be publicly debated in Israel, which is, after all, a parliamentary democracy. This can only be done if the Israeli public knows what is going on at Dimona. So far, crucial nuclear-weapon decisions in Israel have been made by a very small number of political leaders without even a full cabinet discussion. In the absence of any public debate, the authorities can do more or less as they please. As a consequence, Israel's nuclear-weapon developments are probably out of political control.

The fate of Mordechai Vanunu raises a number of crucial questions for the scientific profession. What should a scientist or technician, employed in a military research establishment, do if he or she comes across information, kept secret by the government, that in his or her considered judgement should be made publicly available? If the information concerns activities that are a clear threat to world security, should the international scientific community actively support the scientist or technician if he or she publishes the information and is then put on trial for breaking the country's secrecy laws? Should some institution be set up by the scientific community through which a person can get such information published so that the news media would no longer be the obvious, and possibly the only, outlet for it?

It was obvious to me when talking to Vanunu that, had there been a professional body available to him which would have listened sympathetically to his arguments and considered publishing his information, he would not have gone to a newspaper. His motives for divulging secret information would then have been less open to question and he might not now be facing a long imprisonment.

Given the potential application of some new technologies to weapons of mass destruction, and the wide availability of these technologies, should not the scientific community learn from Vanunu's experience and set up a mechanism to make whistle-blowing a less hazardous business and, therefore, easier to contemplate?

The Vanunu verdict obviously has far-reaching consequences for Vanunu himself but it is also important for Israel, the Arab states and the international community. Israel's political leaders have consistently lied about the nuclear-weapon programme, denying that Israel has any nuclear weapons. Given the current political climate in Israel, the Vanunu verdict will have no effect on that programme.

From Vanunu's information, we know that Israel has produced enough nuclear material for some 150 nuclear and thermonuclear weapons – many more than was previously thought and many more than is necessary to provide an adequate nuclear deterrent. The size and quality of its nuclear arsenal puts Israel in the same nuclear club as China, France and the UK, each of which has a few hundred nuclear weapons, including some thermonuclear weapons.

Although the initial decision to produce nuclear weapons was probably taken for security reasons, there are good reasons for believing that recent nuclear-weapon developments in Israel are the result of the technological momentum of the nuclear programme rather than of deliberate political decisions. Vanunu hoped that once the Israeli public knew the extent of the nuclear-weapon programme a sufficiently strong public opinion would emerge against this development to force the political leaders to bring the programme back under control. This is proving a vain hope.

Vanunu's revelations may even have the opposite effect from the one he intended or from that intended by the Israeli authorities, if they were involved in allowing him to make his revelations. They may hasten the spread of nuclear weapons in the Middle East. Arab leaders can now hardly continue to ignore or explain away Israel's nuclear-weapon capability. They will assume that its development of a sophisticated nuclear arsenal is part of a deliberate policy of acquiring an overwhelming nuclear first-strike capability. Worst-case analysis is always used to judge enemy intentions. Some Arab states are likely to react by accelerating the development of their own nuclear capability. Iraq, Egypt and Libya are planning ambitious nuclear programmes which, if carried through, will enable them to construct nuclear weapons, if they take the political decision to do so.

Only international action can stop the spread of nuclear weapons throughout the Middle East. But, judging by the lack of interest in the Vanunu case, it is unlikely that such action will be taken. If nuclear weapons do spread in the Middle East, there will be real risk that, in a future Arab-Israeli war, they will be used. Vanunu is paying dearly for alerting the world to Israel's nuclear capability. His hope is that, given this information, public opinion will be brought to bear to get rid of the nuclear threat in the Middle East. If it is not, we may all pay even more dearly.

Part I
ISRAEL

2
Israel's Nuclear Requirements

Does Israel have nuclear weapons, or doesn't it? Ask Israeli officials and they will say no. 'Israel will not be the first country to introduce nuclear weapons into the Middle East' is the typical reply. They may add, however, 'Nor can we afford to be the second'! Ask any unofficial person – Israeli or non-Israeli – who has studied the question and you will be told 'Yes, of course Israel has nuclear weapons'.

Who is telling the truth? Are Israeli officials being deliberately misleading? If so, why? Is Israel a fully-fledged nuclear-weapon power? And, if so, for how long has it been one? The first part of this book will present the evidence including a detailed description of Mordechai Vanunu's testimony, so that readers can judge for themselves.

To build nuclear weapons a country has to be able to design them and to produce or acquire the nuclear material needed to fabricate a nuclear explosion. The Appendix provides more detailed information on this. If the design is a relatively sophisticated one, the designers will want to test one of the weapons to make sure that it works. In any case, the military may demand a test before they are prepared to believe that the weapons are reliable enough to put into their arsenal.

The design of nuclear weapons of the types used to destroy Hiroshima and Nagasaki presents no problems to competent nuclear physicists. The information has been available for many years in the open literature. Designers would be confident that the weapons would work as expected and explode with the power predicted. They would see no need for a test; in fact, the Hiroshima bomb, the first of its type, was not tested beforehand.

The nuclear material for weapons – the isotopes plutonium-239 or uranium-235 – is produced in nuclear reactors or in uranium-enrichment plants. All nuclear reactors produce plutonium as a by-product, which can be used as a fuel for nuclear reactors or as the nuclear material to produce nuclear explosions.

The first nuclear reactors were built simply to produce plutonium for nuclear weapons. But most are used nowadays for peaceful purposes – to generate electricity and to produce radioactive isotopes for medical and industrial uses, nuclear physics research, and so on. Nevertheless, a peaceful nuclear programme can have a military component. Any country with nuclear reactors inevitably has the capacity to produce plutonium which could be used to make nuclear weapons. Any country with a significant nuclear programme also has personnel able to produce nuclear weapons. The nuclear scientists and technicians trained to operate even relatively small nuclear reactors can be diverted to a military programme. If a country wants to produce a militarily significant nuclear-weapon force, and to do so clandestinely, it will begin by embarking on a nuclear programme, easily disguised as 'peaceful', leading to the construction of one or more nuclear reactors. This is precisely what Israel did.

ISRAEL'S NUCLEAR PROGRAMME

Israel began its programme to acquire the technology and personnel to produce nuclear materials for weapons almost immediately after the state was born on 14 May 1948. Chaim Weizmann, the world-famous biochemist who was elected first President of Israel, actively encouraged Israeli nuclear scientists. By the end of 1948, a Department of Isotope Research was set up in the Weizmann Institute, the country's main scientific establishment. Separate laboratories worked on applied nuclear physics, electronics, nuclear magnetic resonance and spectroscopy.

In order to pursue this programme seriously Israel knew that eventually it would have to construct a nuclear reactor and a facility to remove chemically the plutonium produced in the reactor. The first step towards acquiring a suitable nuclear reactor was to find the natural uranium needed to fuel it. It soon emerged that Israel had some uranium of its own, mixed with phosphate deposits found in the Negev desert. In 1948, extensive geological teams went to the Negev to determine the extent of these deposits and to assess how much uranium could be extracted from them. The surveys are said to have been carried out by the Israel Defence Ministry (United Nations, 1987). The phosphates were found to contain from 0.01 to 0.2 per cent of uranium and the Negev deposits are estimated to contain between 30,000 and 60,000 tonnes of uranium. Israel has since built three plants to produce phosphoric acid – two in Haifa, each making up to about 15,000 tons of phosphoric acid a year and one in the Negev desert, making about 160,000 tons a year. The uranium currently available from these plants is estimated to be about 100 tons per year (Goldblat, 1985).

Also in 1949 research on the production of heavy water, a material used as a coolant in nuclear reactors, began at the Weizmann Institute. This led to the construction of a heavy-water plant at Rehovot which began operating probably in 1954 (Spector, 1987). The existence of the plant was announced officially in a statement by Foreign Minister Abba Eban on 15 November 1954 to the First Committee of the United Nations on Disarmament.

In 1952, the Ben Gurion government set up the Israel Atomic Energy Commission (IAEC) under the control of the Ministry of Defence. The move was not publicly announced until two years later, by which time the IAEC had signed a secret co-operation agreement with France's Commissariat à Energie Atmoique. This agreement, signed in 1953, was followed by a second one four years later for the building of nuclear facilities at Dimona. Nuclear co-operation continued until the late 1960s (United Nations, 1982).

In the early days of its existence Israel also received important nuclear assistance from the USA. In 1955, an agreement was signed under the Atoms for Peace Programme, established by President Eisenhower. Under this agreement, the US supplied Israel with a small research reactor which was built at the Nuclear Research Centre at Nahal-Soreq and started operating in 1960.

THE DECISION TO MAKE THE BOMB

It is probable that Ben Gurion decided in principle that Israel should develop a nuclear-weapon capability in about 1955. But it was only after the 1956 Suez War, when the United States cut off its arms supplies to Israel, despite Soviet willingness to supply the Arab countries, that determined moves were made to become as self-sufficient as possible in arms production, including nuclear weapons. It was then that the Israeli Government decided to build a clandestine reactor at Dimona to produce plutonium. In the circumstances, the decision seemed amply justified. Nevertheless, it caused controversy amongst those who knew about it and six of the seven members of the IAEC promptly resigned (Van Leeuwen and Soetendorp, 1987). (The objection was on financial rather than moral or strategic grounds; it was felt that the money was more urgently needed for fundamental research.)

The Dimona reactor was kept secret until 1960 when an American U-2 reconnaissance aircraft photographed it. At first, Ben Gurion pretended that the buildings were a 'textile plant' (Pajak, 1982). But his explanation was soon rejected and on 21 December 1960 he admitted to the Knesset that a nuclear reactor was being built at Dimona in the Negev desert but said that its purpose was research and that it would not produce material for nuclear weapons.

Eight years later, however, the government adopted a more open approach to the matter and on 5 October 1968, in the wake of the Six Day War of June 1967, Prime Minister Levi Eshkol announced that Israel 'has the knowledge to make atomic bombs'. Again, fourteen months after the 1973 Yom Kippur War, President Ephraim Katzir, while stating that Israel would not be the first to introduce nuclear weapons into the Middle East, admitted that it had the potential to construct such weapons 'within a reasonable period of time'. And Prime Minister Rabin, interviewed shortly afterwards on British television, repeated that Israel would not be the first to introduce nuclear weapons into the Middle East but added 'We can't afford to be the second either' (Freedman, 1975).

On 25 June 1981, ex-Minister of Defence Moshe Dayan, in an interview with *The New York Times*, said explicitly 'we don't have any atomic bomb now', but added:

> We do have the capacity to produce nuclear weapons, and if the Arabs are willing to introduce nuclear weapons into the Middle East, then Israel should not be too late in having nuclear weapons too.

Earlier, in June 1980, he had said: 'We never said we won't use atomic weapons, we only said we wouldn't be the first to use them' (*Christian Science Monitor*, 18 August 1980).

ASSEMBLING THE MATERIALS FOR THE BOMB

It is clear then that at least from the time of the Suez War, and possibly before that, Israel was determined to assemble the means to build nuclear weapons. What exactly did this involve and how did it do it?

It was from France – the only country willing to supply Israel with military assistance after the Suez War – that a nuclear reactor and a chemical plant to separate plutonium from spent reactor fuel elements were obtained. Both France and Israel have, for the past thirty years, officially denied any exchange of information about nuclear-weapon design and production. Indeed, in his memoirs, President de Gaulle stated that he had effectively stopped Israel's efforts to acquire the facilities 'from which, one bright day, atomic bombs may come' (de Gaulle, 1970). But, in an interview with *The Sunday Times* on 12 October 1986, Professor Francis Perrin, High Commissioner for Atomic Energy from 1951 to 1970, and as such intimately involved with the French nuclear-weapon programme, admitted that the French Government did indeed secretly give Israel details of nuclear-weapon technology. In Perrin's words:

> In 1957 we agreed to build a reactor and a chemical plant for the production of plutonium. We wanted to help Israel. We knew the plutonium could be used for a bomb but we considered also that it could be used for peaceful purposes. It was kept a secret because of the Americans. We had an agreement with them whereby French scientists connected with work on nuclear weapons in Canada (during World War II) could return to France and use their knowledge, but only on condition the secrets would be kept. We considered we could give the secrets to Israel provided they kept them to themselves.

While they worked together French scientists clearly shared important nuclear research results with their Israeli counterparts. It is reported that Israeli scientists were even allowed to observe a nuclear-weapon test at the French test site in the Algerian part of the Sahara desert (Weissman and Krosney, 1981). But in his statement Perrin gave no credit for the nuclear information that flowed from Israel to France. Indeed, the co-operation agreement between the two countries itself came about because France was anxious to find out how Israel was going to produce its heavy water. The technology developed by Israeli scientists in the early 1950s to produce heavy water, probably by hydrogen distillation, was given to the French who built a heavy-water plant at Toulouse; the plant, now closed down, came into operation in 1959 and produced about 1.5 tons of heavy water a year (SIPRI, 1979). The development of France's own nuclear weapons was significantly assisted in this and other ways by fundamental data obtained from Israel's competent nuclear physicists.

Between 1957 and 1959, France and Israel worked closely on nuclear-weapon developments. But, in 1959, following a meeting with Perrin, President de Gaulle decided to stop this co-operation. Nevertheless, the contract for the construction of the reactor and plutonium plant at Dimona was honoured, and the reactor came into operation in 1964.

To start up the Dimona reactor Israel needed a relatively large amount of heavy water and enough natural uranium to fuel the reactor. It could not produce sufficient amounts of either of these materials from its own resources and had to find willing foreign suppliers.

Where Israel's heavy water came from

When the Dimona reactor was started up in 1963 the Renovot heavy-water plant had already been in operation for some ten years. But it is small and probably produces only enough heavy water to top up the reactor when necessary. Israel, therefore, had to buy the heavy water it needed for Dimona from abroad.

The only exporters of heavy water at the time were Norway and the USA. In the late 1950s and early 1960s, Norway was not only producing its own heavy water, but had imported 16 tons from the USA in 1957 which was used in a research reactor (Milhollin, 1986). It therefore had a surplus for export during this period and sold heavy water to several countries, among them Israel. In 1959, the Norwegian Government issued an export licence to the Norsk Hydro company to supply Israel with heavy water, on the basis of an agreement between the two governments which was kept secret until 1979. Between 1959 and 1963 Israel imported 20 tons from Norway (Milhollin, 1987). On 6 February 1987, the Norwegian Foreign Ministry admitted that a further 7 kilograms were sold to the Weizmann Institute in 1964 and 100 kilograms to the Israeli Government in 1968. Between 1959 and 1970, therefore, 21,107 kilograms of heavy water was sold by Norsk Hydro to Israel.

In addition, Israel obtained 3.9 tons of heavy water directly from the USA in 1963 (Milhollin, 1987), the year the Dimona reactor started up and three years after the U-2 plane had photographed the Dimona establishment. The heavy water was supplied under the 1955 bilateral agreement which obligated Israel to use the heavy water for peaceful purposes only and gave the US the right to send inspectors to make sure that this obligation was being fulfilled.

According to Pierre Péan, in his book *Les Deux Bombes*, published in 1982, Israel also received, in 1960, 'a few tons' of heavy water from France. France's own heavy-water production began only in 1967; any exported to Israel must, therefore, have originally come from the USA or Norway or both. Péan maintains that the heavy water was shipped from the French nuclear centre at Saclay directly to the Dimona reactor. He gives much circumstantial detail about who took the heavy water – a senior official of France's Commissariat à l'Energie Atomique – and how it got to Israel – via Bourget and Sicily. He gives no references for his story but claims that the heavy water originally came from Norway.

According to Milhollin's calculations, by 1960 France had imported a total of 36.5 tons of heavy water from the USA (Milhollin, 1986). Some was also imported from Norway but how much is unclear. What is clear is that in 1960 France had more heavy water than it needed for its own reactors (the American supply alone was twice as much as France needed at the time) and that the heavy water was at Saclay.

Péan does not know exactly how much heavy water and French shipped to Israel. But if he is right, France re-exported it illegally, whether it came from the USA or Norway. All deliveries of Norwegian heavy water to France were made on the strict condition that an end-user certificate was produced and an explicit guarantee against re-export given. To import American heavy water, France had to promise to use it only for peaceful purposes and to re-export it only to Euratom countries.

We now know for certain that France secretly built the Dimona reactor and the reprocessing plant in Machon 2 (see Chapter 3). There is, therefore, no reason why it should have baulked at secretly supplying heavy water to allow Israel to start up the Dimona reactor.

Israel's uranium supplies

The Dimona reactor uses natural uranium fuel in the form of solid metal cylindrical rods. According to the United Nations report, *Study on Israeli Nuclear Armament*, between 20 and 25 tons of uranium were initially needed to fuel it. Some 10 tons of this came from Israel's own phosphoric acid plants, as a by-product of phosphate production; the rest was purchased abroad. The Israelis had little trouble in getting hold of uranium in the early 1960s. They were able to obtain it openly on the world market from a number of suppliers, West European and African. Four tons of the uranium for Dimona are reported to have come from France and 10 tons from South Africa (Spector, 1987). Given its past close links with South Africa, there is no reason to doubt that Israel has indeed been supplied with uranium by that country. In addition, over the years it has probably had uranium from Argentina, Belgium, the Central African Republic, Gabon and Niger. Uranium mines in the last three countries were, of course, controlled by France.

Uranium to fuel the Dimona reactor may also have come from a less orthodox source. In 1968, 200 tons of uranium oxide (yellowcake) were sold by the Union Minière du Haut Katanga to an Italian firm and shipped from Antwerp to Genoa. But when the ship carrying it (*The Plumbat*) reached the Mediterranean it was diverted, reportedly to Israel, and the uranium never reached its destination (Davenport et al., 1978).

The history of Israel's efforts to obtain plutonium for nuclear weapons can now be traced with some confidence. But nuclear weapons, it will be recalled, can also be made from enriched uranium in which the proportion of the isotope uranium-235 in natural uranium is considerably increased. There is evidence that Israel obtained stolen enriched uranium and used it in nuclear weapons. Professor Gary Milhollin of the University of Wisconsin School of Law calls the incident – 'the NUMEC-Apollo affair' (Milhollin, 1986).

Between 1962 and 1965, about 100 kilograms of highly enriched uranium disappeared from a factory in Apollo, Pennsylvania owned by the Nuclear Materials and Equipment Corporation (NUMEC). The factory made fuel elements, containing highly enriched uranium, for the nuclear reactors that power US naval submarines and ships. The missing uranium was enough to make at least 6 atomic bombs. The Central Intelligence Agency believes that the uranium went to Israel. In Milhollin's words:

> the head of the NUMEC corporation had close ties to the Israeli military, traces of high-enriched uranium were found at Dimona soon after the material was taken, and Israeli jets began to practice nuclear-weapon bombing runs. In 1968, well before the CIA knew of Israel's plutonium, the CIA told the President that Israel had already made atomic bombs (Milhollin, 1986).

In summary, Israel's nuclear programme is operated by the IAEC, chaired by the Prime Minister, and the National Council for Research and Development. These bodies manage a number of nuclear research institutes and centres. The IAEC, and through it the Israeli Government, operates the Nuclear Research Centres at Nahal-Soreq, attached to Tel Aviv University, and Dimona and the reactors in these centres.

The Dimona Centre is Israel's most advanced nuclear research centre. Whereas the Nahal-Soreq Centre is an open civilian research centre, the Dimona Centre is a top secret military establishment. It works closely with the Ministry of Defence, the Armament Development Authority and the Armament Research and Production Administration which are responsible for nuclear-weapons research and development (Van Leeuwen and Soetendorp, 1987).

Four major university centres train nuclear scientists and engineers, and undertake nuclear research: the Weizmann Institute of Science at Rehovot; Technion, the Israel Institute of Technology at Haifa; the Ben-Gurion University of the Negev at Beersheba; and the Racah Institute of Physics at the Hebrew University, Jerusalem.

DOES ISRAEL NEED TO TEST ITS NUCLEAR WEAPONS?

When assessing the evidence about Israel's nuclear-weapon programmes, the question is often raised as to whether the military and political leaders would be prepared to accept nuclear weapons into the country's arsenal unless and until the designs had been tested in full-scale nuclear tests. In particular, the military may require to know the precise explosive power of any weapons under their control and may demand tests to check that estimated yields can, in practice, be achieved within relatively narrow limits.

Israeli nuclear-weapon designers are competent enough to be confident that they can design nuclear weapons using implosion techniques that would not need full-scale tests. They would also be sure that the weapons would produce explosive yields within their predicted range. Provided that the nuclear material used for fission is weapon-grade uranium or plutonium, the explosive power can be predicted quite precisely, within a narrow range, and the scientists and engineers who built the weapons would be confident that they would explode according to plan.

Whether or not the military will take the word of the nuclear scientists and engineers will depend on their attitude to science and technology. If a significant fraction of the top brass are technically minded it is likely that they will accept the word of the scientists. On this basis, they are very unlikely to demand the testing of ordinary fission nuclear weapons.

But nuclear weapons that include an element of nuclear fusion are a different matter. A scientifically-minded military may be prepared to accept untested boosted nuclear weapons, which essentially consist of ordinary fission weapons with some deuterium and tritium gases fed into the centres of the plutonium spheres. Boosted weapons have the same basic design as ordinary fission weapons – the difference is technical rather than scientific. But the designers of full-scale thermonuclear weapons, using a solid lithium-deuteride fusion component placed outside a fission trigger, are likely to want to test them. Even today, the design of such a weapon is a very complex matter.

The test of a thermonuclear weapon need not involve testing the entire assembly at full explosive power. It will normally be enough to test the fission trigger plus a small section of the fusion component to test that the fusion process is set off. If Israel sets off a nuclear explosion, therefore, it may be taken as an indication that it is testing a thermonuclear-weapon design. The yield of the test may be relatively low; if the scaled-down device produces some fusion, it can be assumed that the full-scale weapon will work effectively.

There is, in fact, some evidence that Israel has performed some nuclear testing. It is important to assess this evidence because the detection of a nuclear test removes all ambiguity about a country's nuclear-weapon programme.

To disguise a nuclear explosion with an explosive power greater than that equivalent to, say, 10,000 tons of TNT, is a difficult task if it is set off below ground. Such an explosion can normally be detected by seismic monitoring equipment operated outside the country in which it takes place. In May 1974 the Indian explosion of a nuclear device with an explosive power of about 12,000 tons of TNT was, for example, so detected. And the Indian explosion, set off at a depth of about 100 metres underground in the Pokharan range of the Rajasthan desert, produced a crater 150 metres in diameter which could be seen from reconnaissance satellites. It is also difficult, but, as will be shown, perhaps less so, to disguise a nuclear explosion in the atmosphere.

On 22 September 1979, an American Vela satellite recorded a double flash of light originating from the South Atlantic-Indian Ocean area. Vela satellites, operated by the US Air Force, specialize in the detection of nuclear explosions in the atmosphere and outer space. Up to that date they had sighted 41 nuclear explosions in 15 years of operation. This was a complete record.

The scientists at the Los Alamos nuclear-weapon laboratory in New Mexico, who operated the Vela satellites, are confident that the signal from the Indian Ocean came from a nuclear explosion. The signal produced in a Vela satellite's equipment by a nuclear explosion in the atmosphere is a very characteristic one – a double pulse in which the heights of the two pulses are in a specific ratio. And this particular satellite's equipment had been calibrated only one week before.

The Los Alamos scientists were not alone in believing that a nuclear weapon had been tested over the Indian Ocean. They were backed up by the US Defense Intelligence Agency, the Central Intelligence Agency, the Naval Research Laboratory and various individuals in the Departments of Energy and State. The CIA was specific about it. The explosion, it said, was a joint South African-Israeli nuclear test. And the US Naval Research Laboratory concluded that the nuclear explosion had occurred over the Indian Ocean, in the vicinity of the Prince Edward and Marion Islands. This conclusion was drawn from an extensive study of all the available evidence – the most thorough study performed by a US government agency, involving some 75 people.

One would have thought that if the main intelligence agencies, the scientists who operated the Vela satellites and the Naval Research Laboratory all agreed that a nuclear test had occurred on 22 September 1979 over the Indian Ocean, this would have been convincing enough. But not so. A scientific panel was set up by the Carter Administration to review the evidence, and came out with the surprising conclusion that the evidence was not conclusive. Its report stated.

> Although we cannot rule out the possibility that this signal was of nuclear origin, the panel considers it more likely that the signal was one of the zoo events, possibly a consequence of the impact of a small meteroid on the satellite (Executive Office of the President, 1980).

A 'zoo event' is one that cannot be adequately explained!

The panel reached its conclusion because this particular signal was bigger than previous Vela signals from nuclear explosions, suggesting that the event occurred, 'close to the satellite rather than near the surface of the earth'. Not a very convincing explanation, particularly considering the evidence, other than that from the Vela satellite, that a nuclear explosion had taken place.

The unsatisfactory nature of the White House panel's work was brought home in a report published on 21 May 1985 by the Washington Office on Africa Educational Fund and written by Ronald Waters (Washington Office on Africa Educational Fund, 1985). This showed that the panel did not receive all the relevant information before it published its findings, and discounted other data brought to its attention which would have led to the conclusion that a nuclear explosion had taken place. The panel, it must be emphasized, did no independent research of its own.

All in all, the White House panel of scientists seems to have been motivated more by political than by scientific considerations. There is no doubt that had it come out with the finding that a nuclear test had taken place, the Carter Administration would have been extremely embarrassed. It would have had to accept the CIA's finding of a joint Israeli-South African test and this, as will be seen, would have had far-reaching repercussions. The issue is, in fact, so important in indicating the intricate international political ramifications of Israel's nuclear activities that it is worth considering in some detail the evidence, which is not widely known, indicating that the White House panel was wrong and that, in fact, a nuclear test had occurred.

The most important piece of new evidence in the Africa Educational Fund's report related to radioactive fall-out. If the event over the Indian Ocean was a nuclear explosion, one would have expected someone, somewhere in the Southern Hemisphere, to have measured some radioactive fall-out from it. And, indeed, soon after the event, a laboratory in New Zealand reported increased levels in radioactivity that could have been caused by an atmospheric nuclear explosion over the Indian Ocean. The winds could have carried the fall-out to New Zealand where it could have been washed out of the atmosphere by rain.

But then a strange thing happened; the New Zealand National Laboratory announced that the fall-out report was wrong. The US Naval Research Laboratory explained that the absence of fall-out could have been caused by the nuclear device being exploded near the surface of the ocean; radioactive fall-out would then not have got into the upper atmosphere and been spread. The Vela scientists at Los Alamos put forward another explanation; they calculated that the nuclear explosion was a relatively small one – equivalent to the explosion of between 2,000 and 3,000 tons of TNT – and suggested that it might have been a neutron bomb.

It turned out, however, that radioactive fall-out was observed, not in New Zealand, but in Australia by a Dr Van Middlesworth. Middlesworth found radioactivity in the thyroids of sheep soon after 22 September 1979. This came from radioactive iodine, which is very efficiently concentrated in the thyroid gland by the normal body processes. Studies of wind and weather patterns confirmed that radioactive fall-out from a nuclear explosion over the southern Indian Ocean could have been carried to Australia – specifically to the states of Victoria and Tasmania where the sheep monitored by Dr Middlesworth were grazing.

Apart from the evidence of radioactive fall-out, the White House panel of scientists also discounted ionospheric evidence provided by the Arecibo Laboratory in Puerto Rico and the Los Alamos weapons laboratory. And the panel did not even consider new ionospheric evidence provided by the US Naval Research Laboratory and radar detections by the US Air Force, all of which indicated a nuclear-weapon test over the Indian Ocean.

The observatory at Arecibo, which has the world's most sophisticated radio telescope, saw an unusual ripple in the ionosphere that could have been caused by a nuclear blast in the atmosphere. The shock wave from a nuclear explosion would travel through the atmosphere and produce disturbances in the ionosphere. The scientists at the laboratory reckoned that the ripple they observed occurred at the same time and in the same general area as the flash detected by the Vela satellite.

Early-warning radars operated by the US Air Force picked up signals on 22 September 1979 of what some analysts believed was a nuclear test. This trio of events – the Vela double flash, the ionospheric ripple, and the radar signals – together with the measurements of radioactivity, are too much to ignore. The suggestion that, although they occurred simultaneously, they were of different, and inexplicable, events is not credible.

For many people, however, the clinching evidence is that a task force of South African warships was conducting a secret exercise at sea on the very night and at the same latitude and longitude as the nuclear explosion is believed to have taken place. South African warships visit that area extremely rarely; indeed, the fact that ships and aircraft avoid the area makes it attractive for a clandestine nuclear test.

The Africa Educational Fund describes how two US spy planes on patrol tried to approach the air space over the area in which the South African warships were sailing but were turned away by the South Africans and forced to land in Australia. This information not only casts further suspicion on the South Africans but also suggests that US intelligence agencies may have had advance knowledge of the planned nuclear test.

There are reasons to believe that, if there was a nuclear test over the Indian Ocean in 1979, efforts were made to render the event ambiguous. The Africa Educational Fund gives two good reasons for suspecting this. First, in 1958, South African scientists had helped monitor a nuclear explosion set off in the atmosphere by the United States in the same area as the 1979 flash. The area (known as the Cape Town Anomaly) had been deliberately chosen because the conditions there allowed the testers to avoid detection; the area has a high background of natural radiation (because it is where the ionized layer of the atmosphere comes closest to the earth) and foreign ships and aircraft steer clear of it. Second, the South African military attaché in Washington had requested the US National Technical Information Service (NTIS) to make 'a computer search of the literature on nuclear explosions and the seismic detection of nuclear explosions, including the flight plans, predicted orbit plans and operations of the Vela satellite –

the only request the NTIS had ever received for that information'.

In spite of these efforts to disguise the event, one would have to be unusually sceptical to deny that the evidence for a nuclear test having taken place overwhelms the evidence against it. Why, then, should the Carter Administration try to cover up the evidence? Even the disclosure of the possibility that a nuclear explosion over the Indian Ocean might have taken place was politically embarrassing to the Administration. Officials tried to suppress the news that a Vela satellite had recorded a double flash; it was leaked four weeks after the event.

Suppressing the news did prevent at least one major diplomatic problem. As the Africa Fund report points out, a South African government delegation was in Washington at the time of the flash to discuss the Antarctica Treaty. This Treaty prohibits nuclear tests in the Antartic region and the flash occurred in Antartica or very close to it.

The evidence linking Israel with the explosion is more circumstantial. A recent account of Israel's military collaboration with South Africa is given by Benjamin. Beit-Hallahmi who says: 'The history of the military alliance has been long and rich in joint projects and collaboration'. It included the production of weapons – such as the Uzi submachine gun – in South Africa under licence. The first Uzis were delivered as early as 1955 and Israel sold South Africa Centurion tanks as early as 1962. (Beit-Hallahmi, 1988). The South Africans have also produced under licence the Israeli Reshef Class fast attack craft (Brzoska and Ohlson, 1986). Mordechai Vanunu spoke of the visits of South African nuclear scientists to the Dimona nuclear research centre. Given the close co-operation over the years between the two countries on the design and production of conventional weapons, similar co-operation is likely on nuclear weapons.

South Africa can produce weapon-grade enriched uranium in its Valindaba plant, which currently has the capacity to produce enough for several nuclear weapons a year. By 1979, however, it probably had not produced enough weapon-grade material for a nuclear weapon. If there was a nuclear test on 22 September 1979 involving a South African naval task force, it is reasonable, given Israel's nuclear collaboration with South Africa, to assume that it was a joint Israeli-South African test using an Israeli-made nuclear explosive. In return for arranging the test, South Africa may well have received Israel's assistance on nuclear-weapon design.

If the Carter Administration had admitted that a nuclear explosion had taken place over the Indian Ocean, it would inevitably have pointed the finger at both Israel and South Africa, with far-reaching consequences. Most importantly, the public and unambiguous admission that Israel had exploded a nuclear device in 1979 would probably have destroyed the 1978 Camp David Accord so carefully nurtured by President Carter himself. The peace treaty between Israel and Egypt, the end product of the negotiations conducted with US participation at Camp David, was signed in Washington on 26 March 1979.

The Camp David Accord and the peace treaty were regarded by the President as his most important foreign policy successes and he was not prepared to damage their prospects. To maintain ambiguity about Israel's nuclear-weapon programme, therefore, suited the Administration's purposes very well. A public admission that Israel had tested a nuclear explosive would have provoked considerable political pressure to stop or reduce US military aid to Israel, and would also have gravely damaged President Carter's Middle East policy. It must also be remembered that 1980 was a presidential election year and Carter did not want to risk losing the votes of Jewish Democrats to Senator Edward Kennedy in the run-up to the Democratic Convention.

An official US admission that South Africa was involved in a nuclear test would also have damaged some elements of President Carter's policy towards that country. In the words of the Africa Educational Fund's report:

> At the time, the US was deeply involved in fashioning the Lancaster House agreement that would transform minority-ruled Rhodesia into independent Zimbabwe. By 1979,

> Carter had softened its anti-apartheid rhetoric, in part to win Pretoria's help in pressuring Rhodesia's Ian Smith to settle. The Carter Administration may have believed that a major US-South Africa confrontation over Pretoria's nuclear weapons program would have scuttled the chances for a successful Zimbabwe independence settlement – another important foreign policy victory the administration needed as it headed into the November 1980 elections.

The fact that no Congressional hearings have been held on the 1979 Indian Ocean event is odd, but this could be a success of the pro-Israel lobby in the United States. In this context, it is noteworthy that Vanunu's revelations about Israel's nuclear-weapon activities have had relatively little political impact in the USA.

In summary, the event of 22 September 1979 over the Indian Ocean was most likely a test of an Israeli nuclear explosive. If so, it was probably a low-yield device, producing an explosion with an explosive power equivalent to that of 2,000 or 3,000 tons of TNT. This may have been a test involving the triggering mechanism for a thermonuclear weapon.

Under current US law no aid can be given to a state, officially classified as a non-nuclear weapon state, that detonates any nuclear explosive device. But the US government will go to great lengths to avoid cutting off aid to Israel, and this is why it will do all it can to avoid having to admit that Israel has tested a nuclear weapon.

DELIVERY SYSTEMS

How could Israel deliver nuclear and thermonuclear weapons? A modern nuclear weapon weighs less than 500 kilograms. The chemical explosives, used to produce the shock waves to compress the plutonium from a slightly greater than critical mass to a supercritical mass, weigh some 200 kilograms. The plutonium normally weighs about 4 kilograms. The other components, the electronics, the casing of the warhead, and so on, account for the rest of the weight.

The American B61 bomb, for example, a modern tactical thermonuclear bomb, weighs a total of about 350 kilograms (and explodes with a power equivalent to that of up to about 500,000 tons of TNT). The American W70 nuclear warhead for the Lance surface-to-surface missile weighs about 200 kilograms; and the W80 nuclear warhead for air- and sea-launched cruise missiles weighs about 120 kilograms (and explodes with a power equivalent to that of up to 200,000 tons of TNT).

The Israeli Air Force operates about 600 combat aircraft, most of which could deliver nuclear weapons. It has been reported that Israeli F-4 Phantom IIs have practised bombing runs of the type used to deliver nuclear weapons and that Israel has tried to acquire from the United States the racks used in the F-4 to carry nuclear weapons. In addition to the F-4, the IAF operates the A-4 Skyhawk, the F-15 Eagle, the F-16 Falcon, all acquired from the USA, and the Kfir, an indigenously designed and constructed aircraft. All these aircraft are capable of delivering nuclear weapons.

The F-4 Phantom is the one most likely to be used by the Israelis to deliver nuclear weapons. It has a maximum range of about 2,100 kilometres and can carry weapons of a total weight of over 7,000 kilograms. When fully loaded its combat radius – i.e., the maximum distance it can travel to a target and return to base at operational altitude and speed with a maximum weapons load – is about 800 kilometres. An American F-4 Phantom, for example, typically carries three nuclear bombs. The F-16 Falcon has a longer combat range – of about 1,500 kilometres. With a weapon load of about 6,000 kilograms, it can carry three nuclear weapons.

In addition to aircraft, the Israelis could deliver nuclear weapons with their 'Jericho' surface-to-surface missiles. The Jericho has an interesting history. The Israelis have always kept the existence of the missile secret. Western sources report that it was developed in collaboration with the French firm Dassault, beginning in 1963. The original missile was based on the design of a French surface-to-surface missile.

The first version of the Jericho – the Jericho I – is thought to have a single-stage, solid-fuelled motor and a maximum range of about 500 kilometres. It was on a mobile version of the Jericho I that nuclear weapons were probably deployed in October 1973.

In June 1987, the authoritative defence journal – the Geneva-based *International Defense Review* – reported that Israel had tested a more sophisticated version – the Jericho II. A missile was fired into the Mediterranean Sea and travelled a distance of 820 kilometres. The maximum range of the Jericho II is projected soon to be about 1,500 kilometres.

The development of the Jericho II is reported to have been started in the early 1970s – probably just after the Yom Kippur War. According to the *International Defense Review*, 'as many as 100 missiles have been deployed, some reportedly in 1985 in the Negev desert and Golan Heights'. According to the publication *Aerospace Daily* of 1 May 1985, Israel has deployed Jericho II missiles with nuclear warheads in the Negev desert since 1981.

This is a very significant development. The Arabs have no comparable surface-to-surface missile. Egypt, Iraq, and Libya have all deployed Soviet-supplied FROG-7 and SCUD-B surface-to-surface missiles. Syria has deployed FROG-7, SCUD-B and also possibly some SS-21 surface-to-surface missiles. The FROG-7 and SCUD-B missiles are very old; they were first deployed by the USSR in 1965. The SS-21 is a modern missile, first deployed in 1978. The FROG-7 has a range of 70 kilometres; the SCUD-8 a range of 300 kilometres; and the SS-21 a range of 120 kilometres.

3
Israel's Nuclear Capability

In 1968, well before the CIA knew details of Israel's plutonium production, the Agency told the US President that Israel had already manufactured nuclear weapons. Whether or not the CIA report was correct, the decision to construct nuclear weapons was probably taken soon after the Six Day War of June 1967. It is reported that Israeli forces captured stocks of Egyptian chemical weapons in the Sinai during the 1967 war. These may have been recommissioned weapons abandoned by the British when they left Egypt in 1952 (SIPRI, 1987). This discovery of Arab weapons of mass destruction may well have hastened the Israeli decision to produce the components for nuclear weapons.

The decision actually to assemble and deploy nuclear weapons was almost certainly made by Prime Minister Golda Meir on 8 October 1973 during the Yom Kippur War (*Time* magazine, 12 April 1976). It is believed that nuclear warheads were then deployed on Jericho I surface-to-surface missiles, probably in reaction to the fear that Syria was about to overrun North Israel. There are also rumours that Israeli nuclear weapons were again deployed on and around 24 October 1973, when Soviet US relations over the Middle East crisis sharply deteriorated. The Soviets may also have sent nuclear warheads to Syria for its Soviet-supplied SCUD surface-to-surface missiles. Prime Minister Meir herself admitted, soon after the war, that there was a moment during the fighting when she feared that Israel might well be beaten and destroyed. It was at such a moment of panic that the decision to deploy nuclear weapons was probably taken.

Nevertheless, as noted in the previous chapter, no Israeli leader has ever admitted unambiguously to having nuclear weapons. Yet for a long time now there have been scraps of evidence to indicate that Israel is in fact a nuclear-weapon power (Spector, 1987), and a great deal of definite information has been given on this point by Mordechai Vanunu.

According to Vanunu, the Israelis are producing at Dimona about 40 kilograms of weapon-grade plutonium a year, and have been doing so for ten, and possibly twenty years. About 4 kilograms is used in each nuclear weapon. Israel has, therefore, produced enough plutonium to construct between 100 and 200 nuclear weapons.

It has also produced about 170 kilograms of lithium-6, which would produce about 220 kilograms of lithium-6-deuteride. Roughly 6 kilograms of lithium-deuteride are needed to construct a thermonuclear weapon (Keith Barham, private communication 1987). Israel may therefore have about 35 thermonuclear weapons.

Before Vanunu's revelations, most experts speculated that Israel had about 20 to 25 nuclear weapons (for example, Spector, 1985). There were suggestions, however, that it had a larger nuclear arsenal; Anthony Cordesmann, for example, on American TV's NBC Nightly News on 30 July 1985, suggested that it had at least 100 nuclear weapons. But Shai Feldman, a leading Israeli strategist and presumably very much in the know, claims rather surprisingly that he does not know whether or not Israel possesses nuclear weapons.

DIMONA'S 'MACHONS'

The Dimona Centre, located about 14 kilometres from Dimona, a town in the Negev desert between Beersheba and Sodom, includes a number of separate blocks, each called a *Machon*. Currently, there are nine Machons in operation; four, Machons 1, 2, 8 and 9, are directly involved in producing materials for nuclear or thermonuclear weapons; the others provide services for these four.

Machon 1, a domed building some 18 metres in diameter, is the nuclear reactor built by France. Machon 2 is the reprocessing plant which removes the plutonium produced in the reactor fuel elements. It also contains a plant to separate the isotope lithium-6 from natural lithium. Machon 8 contains a gas centrifuge plant for the production of enriched uranium. And Machon 9 contains a laser isotope separation facility which can be used to enrich uranium and to increase the proportion of the isotope plutonium-239 in plutonium.

Machon 4 houses a radioactive waste treatment plant. Highly radioactive waste, containing fission products, from the reprocessing plant is sent there for storage. Low-level waste, including radioactively contaminated clothing, tools, equipment, etc., is sealed in canisters and buried on the site. Dimona is not entirely self-sufficient, however. Its electricity normally comes off the ordinary grid although, in an emergency, it can be supplied by generators in Machon 6.

The plutonium-production reactor (Machon 1)

When it started up in 1963, the Dimona reactor was generally believed to have a thermal power output of 26 million watts (MWt). The reactor is moderated by heavy water, the purpose of the moderator being to slow down the neutrons produced in the fission process to low velocities.

Most of the neutrons produced in the fission process travel at relatively high velocities. Natural uranium contains two isotopes – uranium-235 and uranium-238. The more common is uranium-238; natural uranium consists of 99.3 per cent uranium-238 and only 0.7 per cent uranium-235. A neutron can cause a nucleus of uranium-238 to fission only if its velocity exceeds a certain value. But too few of the neutrons produced in the fission process have this critical velocity, the majority are much slower, and a self-sustaining fission chain reaction is not possible using only uranium-238.

Uranium-235, however, will undergo fission when any neutron, even one moving very slowly, collides with it. A chain reaction is, therefore, possible using uranium-235. Plutonium-239 also has the property that it will fission with neutrons of any velocity. And it is this property that makes uranium-235 and plutonium necessary for use as the materials to produce fission in nuclear weapons.

Natural uranium cannot be used on its own to produce a chain reaction because of the large proportion of uranium-238 contained in it. But there is a solution to this problem which allows natural uranium to be used as reactor fuel in a way which does sustain a fission chain reaction. This is to surround the natural uranium fuel with a substance whose nuclei are small in size so that if a fast neutron from fission collides with one of them it will lose a large fraction of its velocity – just as a billiard ball will lose velocity when it collides with another one. The neutron's velocity is thus quickly 'moderated' down to the low velocity at which it can be effectively captured by a uranium-235 nucleus, producing fission, and at which it will have a relatively high probability of avoiding capture by a uranium-238 nucleus.

Some uranium-238 nuclei will capture fission neutrons. The new nucleus formed when this happens will, in general, not undergo fission but will undergo radioactive decay to form plutonium-239. Consequently, when uranium is used as fuel in a nuclear reactor, plutonium steadily accumulates in the fuel elements.

The Dimona reactor is fuelled by natural uranium in the form of rods stacked in the core of the reactor, through which the heavy water circulates. Although its main job is to slow down the fission neutrons, the heavy water is also used as a coolant, to remove heat from the fuel elements in the core to prevent it from over-heating. The heavy water then flows through a heat exchanger where it turns water in a secondary circuit into steam. If the Dimona reactor was used to produce electricity the steam would be used to drive a turbine generator. But the reactor's sole purpose is to produce plutonium for Israel's nuclear-weapon programme. The steam is, therefore, released directly into the environment.

If the Dimona reactor really produced 26 million watts of

thermal power (MWt), it would produce at most 8 kilograms of plutonium a year if run continually. According to Vanunu, however, the reactor is producing 5 kilograms of plutonium a month, for eight months in the year. This means that it must be a lot bigger than is officially admitted. In fact, the thermal power output must be about 150 rather than 26 MWt.

In August 1980, *Foreign Report*, published by *The Economist*, claimed, in an article 'The Middle East's Nuclear Race', that the power of the Dimona reactor was increased in 1980 from 26 to 70 MWt. According to Vanunu, this report is incorrect and no significant increase in power took place while he worked at Dimona, i.e. after 1976. But he does confirm that the power was increased from 26 to 70 MWt before 1976, and that it was again increased, presumably to about 150 MWt. To disperse the extra heat produced, a new large cooling unit was added to the reactor.

The task of making the output of the reactor five or six times bigger without considerably increasing the volume of the core, and hence the size of the reactor building, would have been a difficult one. One way in which this could have been done, although with some difficulty, has been suggested to me by Walter Patterson, an expert in nuclear-power reactors. The Dimona reactor may originally have been moderated with heavy water but cooled using a gas, probably carbon dioxide. There is good reason for believing that this may have been the original design because the French were constructing such a reactor – the EL-4 reactor at the Centre Nucléaire des Monts d'Arrée, near Brennilis Finisterre – at the time they were building Dimona.

The French gas-cooled, heavy-water-moderated reactor went into operation in 1966, with a power output of 250 MWt. It would have been natural to have used the same design for the Dimona reactor. Patterson suggests that the power of the Dimona reactor could have been considerably increased by adding a heavy-water pressure circuit to cool the reactor with heavy water rather than carbon dioxide. With a liquid coolant in contact with the fuel rods, a much greater output could be obtained using the same amount of uranium fuel. In addition, the Israelis may have changed the geometry of the fuel rods somewhat so that they could pack more uranium fuel into the same volume. They could also have increased the power by using enriched rather than natural uranium as fuel, although this is unlikely (the French EL-4 reactor used slightly enriched – about 1.5 per cent – uranium fuel).

Whatever method, or combination of methods, was used to boost the power output of the Dimona reactor in the late 1960s or early 1970s, it would not have been a simple task and its successful conclusion testifies to the competence of Israeli nuclear scientists and engineers. A group capable of such a feat would have had no trouble in designing and constructing efficient nuclear weapons in the late 1960s.

Reprocessing at Dimona

As we have seen, the basic nuclear material for nuclear weapons is plutonium, which is produced in the Dimona reactor and separated from the spent reactor fuel elements in Machon 2. Machon 2 is, therefore, the key building at Dimona; the main role of the other Machons is to support it.

When Vanunu – who studied physics for a year at Ramat Aviv University in Tel Aviv and completed a course lasting several months at Dimona on nuclear physics and chemistry – started his job as a technician at Dimona on 7 August 1977, he went to work in Machon 2. Apart from a short period in 1979, when he worked in Machon 4 (where Dimona's most highly radioactive waste is treated) he worked in Machon 2 until he left Dimona on 27 October 1985.

During this period, he worked in various areas (Units) in which a number of operations are performed, including the removal of aluminium casings from spent reactor fuel elements; the re-purification of heavy water; the removal of uranium from the fluid containing the dissolved fuel elements; the removal of plutonium from the fluid containing the dissolved fuel elements; the production of tritium; and the production of lithium-6. He also visited frequently, and became familiar with, what went on in the Unit in Machon 2 where plutonium and lithium deuteride were machined into nuclear weapon components.

All in all, therefore, Vanunu was familiar with most of the key operations in Machon 2. From the knowledge he picked up from his work experience, he was able to describe the processes. Natural uranium is left in the Dimona reactor for about three months. Fuel elements are then removed in batches and sent to Machon 2 for reprocessing, during which plutonium is separated chemically from unused uranium fuel and the fission products. After three months in the reactor the plutonium produced in the uranium fuel would contain at least 93 per cent of the isotope plutonium-239, ideal fissile material for use in nuclear weapons.

The purpose of Machon 2 – the production of plutonium, tritium and lithium deuteride for use in nuclear and thermonuclear weapons; the purification of heavy water; and the manufacture of nuclear-weapon components – is a closely guarded secret. Only workers with a special pass are allowed inside. Vanunu estimates that of the 2,700 people employed at Dimona, at most 150 know what happens in the building. Only those who work in it realize that it extends deep underground.

The concrete building, which is about 60 metres by 24 metres, has eight floors, one above ground, one at ground level and six (levels one to six) underground. The concrete is several feet thick in some places, protecting it from air attack and shielding workers in some areas from high levels of radiation. Incidentally, the Dimona site itself is protected from air attack by a powerful array of Chapperelle and Hawk surface-to-air missiles.

There are no windows and the building is generally believed to contain just store rooms and workshops. And, indeed, there are store rooms and workshops on the ground floor, with an air filtration unit, offices, shower cubicles and a canteen on the first floor. In fact, the only clue at ground level of the purpose of the building is that one of the four entrances is used for deliveries of spent fuel elements from the reactor. It is in the six secret lower levels that weapon-grade plutonium is removed from these fuel elements.

According to Vanunu, Machon 2 started separating plutonium from spent reactor fuel elements in 1966. By 1972, it was working at full capacity – producing about 1.2 kilograms of plutonium a week for 34 weeks a year, or about 40 kilograms a year. This requires the reprocessing of about 100 tons of spent reactor fuel elements a year.

Although the reprocessing plant is capable of operating virtually all year round, it is closed down each year (from July to November), after eight months of operation, for routine maintenance and cleaning. When in operation, the plant runs continuously around the clock.

While in the reactor, some of the uranium-235 in the natural uranium fuel undergoes fission. The fission process produces fission products – isotopes of elements with atomic numbers ranging from 30 (zinc) to 66 (dysprosium) – and a relatively large amount of fission products is produced. For each megawatt of thermal energy produced, roughly a gram of uranium fuel is consumed a day and the same amount of fission products produced. The bulk of these isotopes are radioactive; when they undergo radioactive decay, they produce radiation and heat. In fact, so much radiation is emitted by the fuel elements that they are dangerous to handle, even with remote-handling equipment. They are therefore stored in a water-filled 'cooling tank' near the reactor for several weeks to 'cool off' before being sent to the reprocessing plant.

During storage, the radioactivity decays by a factor of several thousand, and so does the heat. In addition to making the spent fuel elements easier to handle, the cooling-off period reduces the amount of some fission products present which are a nuisance in reprocessing. Iodine, for instance, is such a volatile element as to be an undesirable contaminant in reprocessing, and iodine-131, a radioactive isotope of iodine with a half-life of just over eight days, is a fission product. But after fourteen weeks in the cooling tank the amount of iodine-131 present in the reactor fuel elements has decreased by a factor of about 4,000, sufficient to remove it as a problem in reprocessing.

The quantity of radioactive isotopes present is measured in a unit called the curie – the amount of radioactive material in which there are 37,000 million disintegrations per second. Relatively non-hazardous concentrations of radioactive contaminants in air and water are, in general, extremely small and are usually stated in millionths of a curie. Even after fourteen weeks, the radioactivity of a gram of fission products is still about 500 curies. Spent reactor fuel elements must therefore be handled with extreme care and with remote-handling equipment to prevent any contamination.

Because of the radioactivity and radiation risks involved, the main chemical operations in the reprocessing plant must be performed by remote control behind thick shielding. The Israelis soon learnt to cope with the remote operation and maintenance needed to enable them to run their reprocessing plant continuously. The process they use is based on an organic solvent, tributyl phosphate (TBP), and is called the Purex (Plutonium-URanium-EXtraction) process. This process dates back to 1954 when the Americans started to use it in their Savannah River reprocessing plant.

The Purex process uses TBP dissolved in a kerosene hydrocarbon as the separating agent. The process depends on the fact that when uranium and plutonium are highly oxidized they are more soluble in the TBP-kerosene solution than they are in an aqueous (water) solution, whereas the fission products are more soluble in a strongly acid aqueous solution than in the organic solution.

Machon 2 is divided into numbered production units. The irradiated reactor fuel elements from Machon 1 are transported in shielded containers to Machon 2, where they enter on the ground floor, in a delivery bay called Unit 10, and are lowered, by crane, to Unit 11 on level three, 10 metres below ground level. Here, the aluminium casing is removed by dissolving it in a tank containing 600 litres of a solution of sodium hydroxide and sodium nitrate. The fluid is very radioactive and is sent to Unit 25 for treatment.

The bare metal fuel element is then dissolved in nitric

acid, forming nitrate solutions of the uranium, plutonium and fission products. Normally, a batch of 140 fuel elements, weighing a total of about 650 kilograms, is dissolved at one time. The nitric acid, heated to 109 degrees centigrade, takes about thirty hours to dissolve the batch of fuel elements.

The concentration of the solution is fixed at 450 grams per litre of uranium and piped by vacuum from the dissolver to level four, the TBP-extraction section or main production hall. The production hall, which contains, in Units 12 to 22, the chemical plant for the separation of plutonium, rises through three levels up from level four into levels three and two. Separation takes place in a number of extraction 'cells' in which organic (TBP) and aqueous solutions, which do not mix, travel counter-currently so that substances more soluble in one solution than in the other are separated out.

The solution from level three (which contains about 30 curies of radioactivity per litre) is fed into a tank in Unit 12 called 'the battery' which contains 17 compartments or 'cells'. The TBP solvent is forced through the cell taking the uranium and plutonium with it. The highly radioactive fission products are removed by a counterflow of aqueous nitric acid and are taken out of the cell. As the process is repeated in successive cells, the solution gets more and more free of radioactive fission products, which are carried by the aqueous solution to a radioactive waste treatment plant in Unit 24.

The solution containing the uranium and plutonium is concentrated in Unit 14 to about 450 grams per litre of uranium with about 130 miligrams per litre of plutonium and sent to Unit 15 which contains the same process as Unit 12 to remove more fission products. In Unit 16, the uranium and plutonium are separated by using a reducing agent, probably ferrous sulphamate, which converts the plutonium into the trivalent state which is not soluble in the TBP solvent, while not affecting the uranium which is soluble in the TBP. The TBP solution containing the plutonium and uranium, which now contains only about a millionth as much radioactivity from fission products as was contained in the solution from Unit 12, is fed into a partitioning column. The plutonium is separated out of the TBP solvent by a downflowing stream of nitric acid containing the reducing agent and taken out of the bottom of the column. An upflowing countercurrent stream of TBP solvent removes the uranium from the top of the column.

Uranium is then removed from the TBP solvent by dilute nitric acid. The plutonium solution and the uranium solution are further purified by similar processes. In passing through each process, the TBP solvent becomes contaminated with fission products, which are removed with an alkaline solution. The Purex process is, however, a very efficient one, and the vast majority – perhaps 99.9 per cent – of the fission products are removed early on in the process.

Vanunu was told that TBP was difficult to obtain and must therefore not be wasted. After use in the reprocessing plant, it is sent to Unit 30 for cleaning and recycling. This strict conservation seems rather surprising because TBP is a derivative of phosphoric acid, of which Israel is a significant producer. The care taken is possibly because TBP is a very toxic material rather than because of its scarcity.

The reprocessing plant in Machon 2 is automated, as indeed it must be because of the huge amounts of radioactivity and high levels of radiation present. The plant is controlled from a control room in level 2, which is on a balcony, called 'the Golda Balcony' after Golda Meir. This gives the controllers a view of the production hall. It was in this control room that Vanunu began his work at Dimona as a controller on 7 August 1977.

When the plutonium nitrate solution leaves the reprocessing centre it contains about 300 milligrams of plutonium per litre. It is sent to Unit 31 in Machon 2 and further concentrated to two grams per litre and stored in a tank holding 400 grams of plutonium in 200 litres.

To produce plutonium metal from the plutonium nitrate solution, the solution is piped from Unit 31 to Unit 36, where separation columns are used to remove the last traces of fission products in aqueous solution. The plutonium is taken off in about 40 litres of an aqueous acid solution containing about 10 grams of plutonium per litre. This solution is sent to Unit 33.

According to Vanunu, at 10 grams per litre the plutonium is suspended in the solution as 'a powder'. Twenty litres of the solution are put in a tank in Unit 33 and heated, he says, with oxalate and hydrogen peroxide for four hours, after which a finer powder is suspended in the solution. After cooling for about eight hours, the liquid is transferred to a large glass bowl in a glove box. At the bottom of the bowl is a glass column into which a saucer-shaped glass dish is placed. Chemicals are added through a pipe, and a plutonium compound precipitates and falls into the glass dish. Air is blown over the precipitate for a few hours to dry it. The material is then sent to Unit 37.

In Unit 37, the plutonium compound is put in an oven for six hours and then hydrogen fluoride gas is passed through the compound for two hours. It is then mixed with calcium and heated to a high temperature in a 'chalk-like pot', which produces a disc of plutonium metal weighing 130 grams. On average, nine of these plutonium discs are produced each week for 34 weeks in the year. A total of 40 kilograms of weapons-grade plutonium is therefore produced per year in Machon 2.

The operations that take place in Units 33 and 37, as described by Vanunu, are presumably the precipitation of plutonium as the oxalate and its thermal degradation to plutonium oxide (PuO_2). The oxide is then converted to plutonium tetrafluoride which is in turn reduced with calcium to plutonium metal.

The plutonium metal discs are sent from Unit 37 to level five, deep underground in Machon 2. Here, in the most secret and secure section of the building, the plutonium is machined into solid spheres, each weighing about 4 kilograms. The spheres are made to exact dimensions and the surface is polished to a mirror-like finish. The lathes and milling machines used for these operations are installed inside large glove boxes. The plutonium metal is worked in an atmosphere of argon, an inert gas, because plutonium tends to ignite spontaneously in air.

Because the plutonium spheres must be kept airtight, they are encased in sealed copper hemispherical shells. These shells, and hemispherical beryllium tampers, two of which are used to surround the plutonium sphere and reflect neutrons back into it, are also manufactured on level five.

These components – the 4-kilogram plutonium-metal sphere, the copper hemispheres and the tamper – are not assembled into a nuclear weapon at Dimona but are taken elsewhere. Vanunu said that once a consignment of components is ready, it is routinely taken at night, in a truck escorted by four private cars carrying armed guards and medical personnel, to a location near Haifa. He believed, but was not sure, that this was a military airfield and that nuclear weapons are assembled there.

The uranium from the reprocessing plant

The TBP solvent containing uranium from Unit 16 is sent to Unit 17 where the uranium is extracted with an aqueous acid solution at a concentration of 70 grams per litre. In Unit 18, the solution is further concentrated to 450 grams of uranium per litre. In Unit 22, the final traces of radioactivity are removed by passing the fluid through a column filled with silica. It is then stored in a 3,000-litre tank in Unit 21 and eventually sent to Machon 3.

In Machon 3, the uranium is reconstituted as fuel for the reactor. In this building, natural uranium metal is produced from yellowcake. It is then sent to Machon 5, where it is machined into rods and coated with aluminium for use as fuel elements in the reactor.

Uranium metal that has become too depleted in uranium-235 to be used as reactor fuel is sent to Machon 10. Some of this depleted uranium, which is the heaviest metal available in relatively large amounts and only mildly radioactive, is machined into munitions – for example, armour-piercing tips for artillery shells. Some of these munitions are used by the Israeli armed forces and some are exported – to Switzerland, for instance.

Disposal of radioactive waste

Apart from a stream of liquid containing plutonium and another stream of liquid containing uranium, Unit 12 produces a third stream containing highly radioactive fission products. This liquid is piped to Unit 24 in Machon 2, built in 1975 to treat highly radioactive waste.

According to Vanunu, the acid stream from Unit 12 is put in a boiler and mixed with sugar to a concentration of 1 kilogram of sugar per 150 litres of liquid. This process breaks down the acid and concentrates the waste to 2,000 curies per litre. The liquid is stored in a 6,500-litre tank for about four years so that some of the radioactivity decays, and it is then sent to Machon 4. Because heat is produced when radioactive isotopes decay, the liquid in the waste-storage tank is cooled with water and air to prevent it from boiling. Unless the liquid is continually mixed and cooled, there is a danger that the tank will explode with catastrophic consequences.

According to Vanunu's description of Dimona's health and safety standards, the safety record is reasonable. Apparently there was a serious accident in 1969 that killed a man in Unit 36, where traces of fission products are removed from the plutonium solution. By mistake, alcohol was used to clean a glass column and the fumes exploded. In a second serious accident, which happened in 1982, hydrogen escaped from an electrolytic cell and exploded. The engineer present was thrown against a wall but was not seriously hurt. There was no radioactive contamination and the unit was back at work in a few days.

Accidents may be few and far between, but Dimona's environmental standards are law. 'Low-level', but still dangerous, radioactive waste, produced mainly in the reprocessing plant, has for many years been mixed with tar, sealed in 200-litre barrels and buried in the desert at disposal sites about a kilometre from Dimona. There is a considerable risk that radioactivity will, in time, leak out of the barrels and contaminate the water table under the Negev desert. According to Vanunu, the Israelis are sufficiently worried about the risk to test regularly for radioactive contamination several kilometres away from the sites.

High-level radioactive waste is stored, apparently permanently and in liquid form, in Machon 4. In an emergency requiring the reprocessing plant to be rapidly drained, the solutions in the plant, which would of course be very radioactive, can be piped into storage tanks kept for this purpose on level six in Machon 2.

The reprocessing plant releases to the atmosphere airborne pollution and radioactive gases. The prevailing winds blow these contaminants across the Jordanian frontier, about 40 kilometres from Dimona. According to Vanunu, especially toxic gases are sometimes released into the atmosphere from the Dimona facilities. But these releases are so controlled that they only happen when the Israeli meteorological office confirms that the wind is blowing towards Jordan!

Production of tritium and lithium

The most interesting information given by Vanunu is about the process used at Dimona to produce lithium-6. Lithium, the lightest known solid, occurs in nature as the mixture of two isotopes – lithium-6 and lithium-7. Most, 92.58 per cent, of natural lithium is lithium-7; only 7.42 per cent is lithium-6. Lithium-6 is needed for nuclear weapons for two purposes: to produce tritium by bombarding lithium-6 with neutrons in a reactor and as a fusion material in a thermonuclear weapon, where it is compounded with deuterium, an isotope of hydrogen, to produce lithium-6 deuteride. But for these purposes lithium-6 has to be separated from lithium-7, its sister isotope in natural lithium.

In 1977, the Israelis built a pilot plant to enrich lithium-6. After solving some initial problems, a full-scale production plant was built in Unit 95 in Machon 2; by 1984, it was in full production. But after the plant had been in operation for three years, the Israelis decided that they had enough lithium-6 and suspended production. In the first year the plant produced about 36 kilograms of lithium-6; in the next two years it produced about 130 kilograms, for a total over the three years of about 170 kilograms.

Lithium-6 is enriched in Unit 95 in an apparatus that consists of six columns, each 13 metres long, housed in a disused lift shaft. The lithium is imported from the United States by commercial firms.

The method used first involved the production of an amalgam of lithium and mercury. According to Vanunu, this was done by electrolysis. Lithium hydroxide was then passed through the lithium-mercury amalgam and separation of lithium-6 from lithium-7 occurred by exchange between the amalgam and the aqueous solution of lithium hydroxide. The lithium-7 was concentrated in the amalgam phase and the lithium hydroxide because enriched in lithium-6.

By repeating the process through each of the six columns, the proportion of lithium-6 was, according to Vanunu increased from the natural proportion (7.42 per cent) to about 85 per cent. The depleted lithium was separated in Unit 99 from the amalgam as a solid and returned to its owners, although the proportion of lithium-6 had been reduced to about 5 per cent!

After succeeding in producing lithium-6, the Israelis started to produce tritium. Small rods of lithium-6 and aluminium were irradiated with neutrons in the core of the reactor. The irradiation produced tritium, helium and hydrogen.

The irradiated rods were taken from the reactor to Unit 93 in Machon 2, where they were heated in an oven to about 625 degrees Centigrade. The aluminium melted and the tritium, helium and hydrogen gases were given off. Tritium and hydrogen were separated by passing them through a column of palladium asbestos. Helium and tritium were separated by passing the mixture through a column containing palladium and mercury.

The tritium gas was stored by passing it through powdered uranium which absorbed the tritium. When the tritium was required, the uranium was heated and the tritium came off.

According to Vanunu, the tritium and lithium production units were regarded by the authorities as the most prestigious at Dimona. They were always proudly displayed to visiting dignitaries, particularly during visits of Israeli Prime Ministers and Ministers of Defence.

Uranium enrichment

Uranium-235 is an alternative fissile material to plutonium-239. For use in nuclear weapons, the amount of uranium-235 in uranium must be increased (i.e. enriched) from the natural value of 0.7 per cent to over 50 per cent, and preferably to over 90 per cent. According to Vanunu, a secret unit in Machon 8 has been enriching uranium using gas centrifuges since 1979 or 1980. And, in 1981 in Machon 9, the Israelis began using laser beams to separate uranium isotopes. He did not know, however, how much uranium was being produced by these methods but claimed that the laser-enrichment unit was being expanded to production scale when he left Dimona in 1985. Presumably, this plant is producing several kilograms of enriched uranium a year.

Israeli scientists are, in fact, leaders in the sophisticated field of laser isotope enrichment techniques. In the magazine *Science* of 22 March 1974, Robert Gillette reported that Isaiah Nebenzahl, a physicist with Israel's Ministry of Defence, admitted that he and another Israeli scientist, Menahem Levin, had demonstrated the feasibility of laser enrichment. They were not far behind American scientists in developing a technology that greatly reduces the cost and complexity of enriching uranium for use in nuclear weapons. This is yet another indication of the competence of Israeli scientists.

Summary of processes at Dimona

In summary, the Dimona nuclear centre is divided into nine independent production units, each occupying a separate building on the site of the centre:–

Machon 1. The 150 MWt plutonium-production reactor.

Machon 2. The plutonium-separation plant; the lithium-6 separation plant; the tritium production plant; the manufacture of plutonium metal spheres, berylium tamper and other nuclear-weapon components; and the purification of heavy water.
Machon 3. Production of natural uranium metal from yellowcake (uranium oxide – U_3O_8; reconstitution of unused uranium separated in Machon 2 from spent reactor fuel elements; uranium metal sent to Machon 5; and lithium-6 solidified for production of tritium in the reactor.
Machon 4. Plant to treat radioactive waste. Highly radioactive waste stored permanently as a liquid in tanks; low-level radioactive waste mixed with tar buried in desert in large cans.
Machon 5. Reactor fuel fabrication plant in which solid uranium rods are encased in aluminium.
Machon 6. Supplies services – electricity, chemicals, steam, etc. – to other Machons; houses emergency electricity generators.
Machon 7. Probably does not exist.
Machon 8. Gas centrifuge plant to produce enriched uranium; laboratory for testing the purity of samples from Machon 2 and experimenting on new processes.
Machon 9. Experiments on laser separation of the isotopes of uranium.
Machon 10. Depleted uranium – i.e. uranium containing too little uranium-235 to be usable as reactor fuel – made into armoured-piercing tips for artillery shells.

Of these Machons, Machons 1 and 2 are the crucial ones. It is in them that plutonium, the basic fissile material for nuclear weapons, is produced, separated, purified, and, in the metal form, made into spheres for assembly into nuclear weapons.

Machon 2 has eight floors – two above ground and six below ground:–
First floor. Administrative offices, showers, canteen and air filtration unit. Unit 40 provides some services for the Machon – including water cooling, vacuum production, and the preparation of acid and alkaline solutions.
Ground floor. Irradiated fuel elements are unloaded at one

entrance and lowered to level three, below ground; store rooms and workshops in which apparatus for the Machon is made.

Level one underground. Supplies services – electricity, steam, chemicals, and so on – by pipes to other levels. Some of these services come from Machon 6.

Level two underground. Main control room (about 30 metres long) for automated reprocessing plant, with 'Golda Balcony'.

Level three underground. Spent reactor fuel elements received from ground floor, their aluminium cases removed and the uranium dissolved in nitric acid; the extremely corrosive and radioactive solution transferred by vacuum through special pipes to level four; laboratories check the concentration and purity of the chemicals used in the reprocessing plant.

Level four underground. The plant in which the main chemical processes for the separation of plutonium and uranium from the spent reactor fuel elements are performed rises from this level; a small control room monitors both the production of tritium and the extraction of plutonium from the nitrate solution and its conversion into pure metal.

Level five underground. Plutonium, lithium deuteride and beryllium are machined, in large glove boxes, into nuclear-weapon components; the lithium-6 separation plant rises up to level two in a disused lift shaft.

Level six underground. Tanks for use only in an emergency, to contain the liquids in the reprocessing plant if the plant has to be rapidly drained.

Machon 2 is divided into a number of smaller production units. The most important of these are:–

Unit 10. Receives trucks carrying containers of fuel rods that have been irradiated in the reactor. The containers are lowered by crane to Unit 11.

Unit 11. Aluminium casings on fuel elements dissolved in a solution of sodium hydroxide and sodium nitrate, which is then highly radioactive and sent to Unit 25 for treatment. The fuel elements (typically 450 kilograms of them) are dissolved in nitric acid, the solution is adjusted to contain

450 grams per litre of uranium in 1.5-normal nitric acid and transferred through pipes by vacuum at a rate of about 30 litres per hour to Unit 12.

Unit 12. Here the Purex process – involving the solvent TBP – begins and the radioactive fission products are removed in an aqueous acid solution and the highly radioactive liquid sent to Unit 24 for treatment. The uranium and plutonium are carried in the TBP solution to Unit 13.

Units 13, 14 and 15. Traces of fission products are removed and the fluid concentrated to 450 grams per litre of uranium and 140 milligrams per litre of plutonium.

Unit 16. The uranium and plutonium are separated – the plutonium is extracted in an aqueous nitric acid solution and sent to Unit 31; the uranium is carried off in the TBP solvent to Unit 17.

Unit 17. The uranium is extracted with dilute nitric acid and, at a concentration of 90 grams per litre, sent to Unit 18.

Unit 18. The uranium solution is concentrated to 450 grams per litre and transferred to Unit 21 via Unit 22.

Unit 22. The fluid is passed through a column containing silica to remove the last traces of fission products and stored in Unit 21 in a large tank. It is eventually sent to Machon 3 where uranium is removed from the nitrate solution, converted into metal, and ultimately may be recycled as fuel in the reactor.

Unit 24. Treatment of highly radioactive waste solution of fission products from Unit 12, by use of sugar to break down the acid solution and concentrate the waste solution. It is then stored for a few years to allow the radioactivity to decay somewhat before being transferred to Machon 4 for permanent storage.

Unit 30. TBP cleaned and decontaminated for further use.

Unit 31. Plutonium nitrate solution concentrated to about 2 grams per litre and stored in 200-litre cans. Final traces of fission products are removed in Unit 36 and the solution sent to Unit 33.

Unit 33. Oxalic acid is added to the plutonium nitrate solution to precipitate plutonium oxalate, which is removed from the solution and dried. The dry powder is probably calcined in an argon atmosphere at a high temperature to convert it to plutonium oxide (PuO_2). The oxide powder goes to Unit 37.

Unit 37. The plutonium oxide is converted into plutonium tetrafluoride and then into plutonium metal in the form of 130-gram discs, which are sent to level five where they are machined into 4-kilogram spheres for use in nuclear weapons.

HEAVY WATER – THE ADDITIONAL EVIDENCE

Vanunu's statements that the power output of the Dimona reactor, the crucial element in Israel's nuclear-weapon programme, was increased from 26 to 70 MWt and then again to over 100 MWt, are supported by what we know of Israel's supplies of heavy water. Assuming that the reactor was built on the pattern of the French EL-4 reactor and initially cooled with a gas and moderated with heavy water, the amount of heavy water originally needed would have been about 10 tons. The 20 tons we know for certain was originally imported from Norway was therefore twice as much as was needed.

If Israel added the 4 tons of heavy water obtained from the USA under the Atoms for Peace programme and got, say, 4 tons from France, to make up a total of 28 tons, it would have enough heavy water for a 70 MWt reactor. Israel's probable heavy-water inventory therefore supports Vanunu's story that the output of the Dimona reactor was increased to 70 MWt.

Incidentally, Péan maintains that original cooling circuits were built three times larger than was needed for a 26 MWt reactor, much to the puzzlement of the French engineers on the Dimona site. It is, of course, possible that the power output was about 70 MWt right from the beginning.

We know that, to produce 40 kilograms of plutonium a year, the power output of the Dimona reactor must have been increased again to much more than 70 MWt – in fact, to about twice as much. If this increase in power was achieved by modifying the reactor to use heavy water as the coolant as well as the moderator, Israel would have had to import an extra 60 tons or so of heavy water.

Where could all this heavy water have come from? In July 1970, the Norwegian Department of Commerce issued a second export licence allowing Norsk Hydro to export one ton of heavy water to Israel, and at the same time asked the Norwegian Foreign Office to consider an application for a licence to export another 4 tons of heavy water to Israel. The application was rejected 'in view of the political developments in the area'. Given this refusal on the part of the Foreign Office to sanction the export of more heavy water to Israel, it would seem that the export of one ton in 1970 slipped through the system by mistake.

As we saw in Chapter 2, the total amount of heavy water sold to Israel by Norsk Hydro was 21,107 kilograms. It is clear therefore that Norway did not supply Israel with the extra heavy water it needed. But France almost certainly did.

By 1970, France had imported about 200 tons of heavy water from the United States. And France's own heavy-water production plant began operating at Mazingarbe in 1964, producing about 26 tons of heavy water a year. France could, therefore, have supplied Israel with the heavy water it needed to boost Dimona's power from 70 up to 150 MWt.

4
What are Israel's Nuclear Weapons For?

Israel's political leaders have never admitted or denied that Israel has nuclear weapons. Nuclear ambiguity has served Israeli purposes for a number of reasons. Firm knowledge that Israel has a nuclear arsenal much larger than that needed for last-ditch nuclear deterrence would encourage the Arab states to acquire nuclear weapons; would make a pre-emptive Arab attack against Israel's nuclear-weapon sites and stores, nuclear command and control centres and other nuclear-related targets more likely; would encourage the Soviet Union to give its Middle Eastern allies firm guarantees of nuclear protection against Israeli nuclear attack by threatening Israel with nuclear retaliation in the event of a nuclear attack on its neighbours; and, last but not least, would greatly complicate Israel's relations with the United States. In particular, Congress might well be less willing to continue to supply Israel with sophisticated conventional weapons.

Nevertheless, some influential Israeli strategists argue against the government's policy of nuclear ambiguity. One such expert, Shai Feldman, a firm believer that nuclear weapons provide their owners with an effective deterrent, argues that if Israel wishes to maximize its deterrent capability it should 'acquire the capability and adopt a doctrine of overt, disclosed, and explicit strategic nuclear deterrence' (Feldman, 1982). But he also argues that to reduce the political costs of adopting such an overt nuclear policy, it should 'attempt to postpone the adoption of overt nuclear deterrence until a few more states have developed a nuclear capability. It would be still more desirable for

Israeli disclosure to follow the acquisition of a nuclear capability by at least one Arab or Muslim state.' It should be emphasized that Feldman couples the adoption of an overt nuclear deterrent with 'a flexible political posture with respect to the territories it has held since the 1967 war'. In other words, the deterrent should not be aimed at maintaining Israel's occupation of the West Bank, the Golan Heights, and the Gaza Strip.

There are differing views within Israel about the wisdom of possessing nuclear weapons, some of which are surprising. For example, several otherwise dovish factions of the Labour Party are in favour because they believe that Israel can give up the West Bank if it has the security of nuclear weapons. And some hawkish factions, headed, for example, by former Defence Minister Ariel Sharon, are against nuclear weapons because they want to increase Israel's conventional military strength.

The military establishment and its supporters are, of course, mainly influenced by the relative military strength of Israel's adversaries, which affects their thinking about the need for nuclear weapons. Syria, Iraq, and Jordan alone dispose of wartime armies totalling some 1,800,000 soldiers, 10,000 main battle tanks and 1,342 fighter aircraft. Facing them is Israel's wartime strength of about 440,000 soldiers, 4,000 tanks and 662 fighter aircraft.

For purposes of comparison, these Arab armies are roughly the same size as NATO's total active ground forces plus its ground force reserves deployed in Central Europe (i.e., the NATO forces deployed in West Germany, Belgium, the Netherlands, and Luxembourg). The number of Arab main battle tanks is about the same as the total number of NATO's main battle tanks deployed in Central Europe. And the number of Arab fighter aircraft is four times as many as NATO has in Central Europe.

If Egypt's military strength – 320,000 soldiers, 2,250 main battle tanks, and 160 fighter aircraft – is added, Israel is even more outnumbered. In the Arab-Israeli wars so far, Arab military operations have been very badly co-ordinated, and this has greatly reduced their effectiveness. Israeli strategists must, however, assume that in the future the Arabs may get their military act together and present, to say the least, a serious threat.

This threat will be the basis of the arguments of those groups within Israel's military establishment who argue for the deployment of tactical nuclear weapons to counter a massed Arab tank attack. The need to reduce collateral damage and limit radioactive fall-out, because of the smallness of Israel's territory and the vulnerability of its population, will be used by those advocating the deployment of enhanced radiation warheads (neutron bombs) as anti-tank weapons. In this context, Ariel Sharon is right; it is much more cost-effective to deploy conventional anti-tank guns, missiles and mines against tanks than to deploy neutron bombs.

ISRAEL'S TECHNOLOGICAL SUPERIORITY – WILL IT LAST?

Israel's military inferiority has so far been offset by its technological superiority and the better fighting spirit of its troops. The tactics, operational skills and motivation of the troops are perhaps the most important factors determining the performance of armed forces in wartime. And morale inevitably plays a critical role. The success of a military policy will be determined by whether or not the armed forces understand their function and are convinced of its usefulness.

Israel's superiority in military technology, and in tactics and operational skills, was well demonstrated by the Israeli operation on 9 June 1982, during the Lebanon War, in which the Israeli Air Force destroyed Syria's sophisticated surface-to-air missile complex in the Bekaa Valley. The operation began when the Israelis sent a wave of remotely-piloted vehicles or RPVs (automated pilotless aircraft) against the Syrian air-defence system in the Bekaa. The Syrians switched on their radars to detect and track the RPVs and attacked them with surface-to-air missiles. But Israeli aircraft, flying some distance away, included some F-4 Phantoms carrying Shrike and Standard air-to-surface missiles that could detect the rays emitted by the radars and home in on them. Other Israeli aircraft involved were F-16 Falcons, armed with autonomous stand-off missiles capable of being launched from a distance and finding their targets without further instructions.

The Israeli raid was a 'surgical' strike, taking just a few minutes. Some thirty surface-to-air missile sites, the bulk of the Syrian air-defence system, were destroyed. Throughout the attack, Israeli RPVs, carrying television cameras, circled the area and transmitted continuous coverage of the events to the Israeli commander in his ground-based command centre far from the battle.

The Israeli pilots were provided with continuous information from several E-2C spy aircraft flying up and down off the coast of Lebanon. E-2Cs can pick up enemy aircraft 350 kilometres away and continuously track them; they were able to monitor the airspace over the Bekaa Valley and warn the Israeli pilots of the approach of any Syrian fighters. An Israeli Air Force Boeing 707 aircraft, packed with electronic intelligence equipment, monitored Syrian radars. And CH-53 Stallion helicopters carried jamming equipment to wipe out voice and other communications between Syrian fighter pilots and their ground controllers. So good was the electronic intelligence that the Israeli pilots, in addition to destroying the Syrian surface-to-air missile sites, were able to shoot down as many as 85 Syrian MiG fighters with the loss of only two Israeli aircraft.

The Bekaa Valley action is one example of many of Israeli prowess in electronic warfare, a prowess that the Arabs cannot yet match. Combined with a superior fighting spirit and high morale, Israel's technological lead has served it well.

But how long will these advantages last? It is, of course, impossible to predict when the psychological state of a society will change. Some of the Israelis who believe that there will be other Arab-Israeli wars and fear that Israel will lose its technological lead and its will to win against great odds tend to argue that it should have a large and high-quality nuclear force. But many who argue in this way do not think through the consequences of an Israeli nuclear force and to what use such a force could be put.

AN ISRAELI PERSPECTIVE

From the perspective of most ordinary Israelis, nuclear weapons are an insurance against the day when Israel loses its conventional military technological superiority over the Arabs and needs a deterrent against an Arab attack with chemical (or biological) weapons. Some believe that Israel's survival depends on its having all conceivable weapons. And others believe that it needs nuclear weapons as a deterrent against their use by the PLO or some other sub-national group.

For most Israelis these views are likely to be gut reactions rather than carefully thought-out opinions. The fact that the use of nuclear weapons in the region would be so damaging as to be suicidal and, therefore, totally irrational, does not affect the conclusion that they draw – that Israel should have nuclear weapons. For some, of course, it would be justifiable to use nuclear weapons in revenge if Israel was about to be annihilated.

But to understand the strength of feeling behind Israeli attitudes to nuclear weapons it is necessary also to appreciate how thoughtful and influential, but not hawkish, Israelis see the geopolitical situation in the Middle East. I have recently spent some time discussing this with a number of such people. The views of Shalheveth Freier, a senior scientist and former Director of Israel's Atomic Energy Commission and now an Israeli representative at the United Nations, are typical. He has described them to me as follows.

Regional agreements in the Middle East are unreliable. In the past 26 years, Egypt and Syria joined together in the United Arab Republic, which soon fell apart; a union of Syria and Iraq came to nothing and the two countries are now enemies; a similar plan for Syria and Libya miscarried; and many agreements between the PLO and the Lebanese Government have broken down. The Saudi Arabian regime and the Sheikhdoms in the Gulf are unstable. Libya practises subversion in the Sudan and Chad. For these reasons, Israelis are reluctant to rely on formal agreements, at their face value, with their hostile neighbours. Freier, like most Israelis, believes that Israel must rely on itself for its own defence. It must not bank even on the United States.

Israel has 'a very small margin of error'. Some 4 million Israelis see themselves surrounded by 90 million actively or potentially hostile people. This excludes Egypt with which Israel is now at peace, 'the durability of which has not yet been tested'. Israelis feel permanently under siege.

> The countries which until recently avowed the dismantlement of Israel cover an area of 22 million square kilometres as against the 22 thousand square kilometres (about the area of Wales) administered by Israel, the pre-1967 area of which fits into New York City.

As regards wealth, the gross domestic product of Israel ($22,160 million in 1986) is a mere quarter of that of Saudi Arabia alone ($82,440 million in 1986). Money talks, and the disparity in wealth, Israelis believe, influences the attitude of the industrialized countries towards Israel and the Arab countries.

> Against this background, Israel's margin of permissible error remains extremely small, and much of Israel's preemptory military action, when it senses danger, should be understood on this basis. Such small margins of permissible error count heavily in Israeli perceptions.

An important part of the Israeli perspective is that they do not expect fair treatment from the international community. Older Israelis learnt this lesson from their pre-1948 experiences. Israelis born after 1948 (now the majority) have learnt it from, for example, the attitude of the United Nations and its agencies which spend much time in moral condemnations of Israel. The European countries, in their pro-Arab policies, are seen to be sacrificing principles to political or economic expediency. Feelings of 'automatic and unconditional ostracism on the international scene cannot but stiffen the policies of an Israeli government'.

This then is the way even thoughtful Israelis view their predicament. Most believe that if they ever become the underdog, they will go under. They must, therefore, be as powerful militarily as they can possibly be. To be less than this, they believe, is to invite catastrophe. And nuclear weapons are seen by most of them as an important element of military power.

This ingrained attitude towards nuclear weapons explains why there has been virtually no nuclear debate in Israel at a time when nuclear issues are energetically discussed in other democracies. The revelations of Mordechai Vanunu and his trial have not led to any domestic questioning of Israel's nuclear-weapon programme and policy. Few ask why Israel needs so many nuclear weapons and long-range missile delivery systems.

This surprising acquiescence on the part of such a democratic society indicates a general acceptance that, for Israel's security, all types of weapons are necessary. No debate is necessary. But what the Israelis have not yet realized is that, in the absence of parliamentary and public scrutiny, the momentum of nuclear military technology soon takes over and politicians lose control of nuclear policy. Israel's policy of nuclear ambiguity greatly encourages this process, because it can only work in secrecy. Public debate destroys ambiguity.

That Israel has real and serious security problems is undeniable. With a population of 4.3 million, it is confronted by heavily-armed Arab states with a total population of some 140 million. Most Israelis believe, rightly or wrongly, that the destruction of the Israeli state remains the ambition of at least some Arabs. This belief was much stronger in the 1960s when the Ben Gurion government decided to embark on a nuclear-weapon programme. Security was the main reason for the initial political decision to go nuclear.

In actual fact, the United States would never allow Israel to be eliminated; if it did, its other alliances, including NATO, would collapse. But Israelis refuse to rely on this *de facto* security guarantee. The memory that no country was prepared to do much to help when Hitler murdered

6 million Jews makes them doubt that any country would come to their aid if they were being pushed into the sea. And, given the history of the Jews, who can blame them?

Because they count on no other country for help in an emergency, the Israelis want to be as self-sufficient as possible in weapon production, including nuclear weapons. The drive for self-sufficiency is enhanced by such factors as the provision by both the USA and the USSR of sophisticated weapons to the Arab states so that the technological gap between Israel and its enemies has narrowed ominously. Until recently this gap had been a great comfort to Israel.

The knowledge that Iraq is producing significant amounts of chemical weapons, and has demonstrated that it is prepared to use them, and the suspicion that Syria and Egypt may also have chemical weapons, are other reasons for Israel to wonder how long it can maintain its military technological edge. In this situation, some Israelis are bound to argue for more and better nuclear weapons.

Chemical weapons such as nerve gases are, like nuclear weapons, weapons of mass destruction. A large-scale Arab chemical-weapon attack on Israel, particularly one that included civilian populations, would probably be countered by an Israeli nuclear attack against Arab targets. Vanunu said that he had been told that Israel was itself producing chemical (as well as nuclear) weapons at Dimona but that he had no direct knowledge of this.

Apart from chemical weapons, Israeli strategists must take into account the possible acquisition by Arab countries of other weapons of mass destruction, such as biological and radiological weapons. Military technology is also making available very destructive conventional weapons. Fragmentation warheads delivered, for example, by multiple rocket launching systems, fuel air explosives, and so on, are as destructive locally as low-yield nuclear weapons. Very powerful conventional warheads could be delivered on cities by aircraft or missiles. Nuclear weapons may be seen as a way of deterring the use against Israeli populations of non-nuclear weapons of mass destruction.

NUCLEAR DETERRENCE

Most commentators assume, however, that Israel's nuclear weapons are to provide a last-ditch deterrent – to prevent the Arabs *in extremis* from making those military moves that would threaten Israel's very existence. Not surprisingly, a principal Israeli fear is of defeat in a future conventional war followed by the occupation of the Israeli homeland and the massacre of the Jewish population. Robert Harkavy adds what he calls the 'withdrawal scenario' – that Israel's nuclear weapons could deter a massacre of the Israeli population, when its evacuation from the region might be feasible if there was enough time. The threat to use nuclear weapons might buy this time (Harkavy, 1986).

Such justifications for Israel's nuclear force do not account for the size and quality of its nuclear weapons. A 'last-ditch' nuclear deterrent would be adequately provided by ordinary fission nuclear weapons targeted on the major Arab cities – in Syria, Iraq, Egypt, Saudi Arabia and, possibly, Jordan and Libya. A dozen or so nuclear weapons with explosive powers equivalent to that of about 20,000 tons of TNT (the size of the nuclear weapon that destroyed Nagasaki) would do this job. Thermonuclear weapons would not be needed for such a strategy. No Arab city is big enough to 'justify' a thermonuclear weapon.

Why, then, has Israel opted for a relatively large and sophisticated nuclear force? The most likely explanation seems to be that the technological momentum of the nuclear-weapon programme has taken over and become unstoppable. Israel has had to form a team of nuclear scientists and technologists to operate its nuclear reactors and its reprocessing plant, and to design, develop and produce nuclear weapons. These professionals will want to design and produce increasingly sophisticated nuclear weapons just to convince themselves that they can do so and for the sheer satisfaction of it.

If this is really what has happened in Israel, Israeli nuclear-weaponeers are no different from their counterparts in other nuclear-weapon powers. There is, after all, no rational military or political reason for any country to produce high-yield thermonuclear weapons. As American nuclear-weapon scientist Robert Oppenheimer pointed out in the late 1950s, boosted nuclear fission weapons (i.e. fission nuclear weapons containing deuterium and tritium gases in their centres) are big enough for any conceivable targets, including large cities. Bigger yields are simply not justified. But Oppenheimer was silenced and the American nuclear-weapon community, closely followed by the Soviet one, went on to produce megaton-yield thermonuclear weapons. The United Kingdom, France and China all followed in their footsteps. Israel seems to have done the same.

Although technological momentum may have been the main driving force behind Israeli nuclear-weapon developments, the security establishment was probably by and large sympathetic to requests from the Dimona staff to produce the materials and make the components for thermonuclear weapons, particularly during the 1970s. In the mid-1970s, for example, tension was high and there was a fear that war with Syria was imminent. Moreover, moves by Pakistan to produce enriched uranium, suitable for use in nuclear weapons, were another powerful argument for those in Israel wanting to continue developing nuclear weapons.

At the same time, Israeli weapon systems that could be used to deliver nuclear weapons are being continually improved. The combination of sophisticated nuclear warheads and new delivery systems may well be changing Israel's nuclear policy from nuclear deterrence to nuclear war-fighting. This change may not reflect the current wishes of the political leadership but be brought on by the momentum of military technology.

The nuclear policies of the advanced nuclear-weapon powers, particularly the USA and the USSR, have changed dramatically because of technological advances in nuclear weapons and their supporting technologies. It is useful to describe in some detail the experiences of these powers because similar changes are almost certain to occur in

Israel's nuclear policy. In fact, they are virtually inevitable if significant resources are given to a competent team of nuclear scientists and engineers to develop nuclear weapons and then continually modernize them.

The targets at which nuclear weapons are aimed are generally determined by the accuracy with which they can be delivered. Inaccurate nuclear weapons are seen to be useful for deterrence, by threatening an enemy with unacceptable death and destruction; accurate nuclear weapons are seen as more useful for fighting a nuclear war than for deterring one by assured destruction.

Generally speaking, military deterrence – nuclear or non-nuclear – depends on the theory that if your potential enemy knows that he will suffer unacceptable damage if he attacks you, or that the attack is very likely to fail, he will not attack in the first place. The origin of *nuclear* deterrence dates back to the late 1940s and early 1950s, when the United States deployed a large nuclear arsenal and needed to evolve a doctrine to justify it. The doctrine chosen was nuclear deterrence and a number of defence intellectuals developed a theory to support it.

After the end of World War II, the Americans claimed that they needed nuclear weapons to counter a perceived threat from considerably superior Soviet conventional forces. The argument was that the Soviets would not attack Western Europe with their conventional weapons if they knew that their cities would, in retaliation, be destroyed with nuclear weapons.

This strategy of massive retaliation changed when the Soviet Union developed a nuclear arsenal comparable with that of the United States. It was all very well to threaten the Soviets with genocide when they could not reply in kind. But when they could, NATO changed its tune and took on a policy flexible response. This is based on a 'ladder of escalation' starting with the use, if necessary, of low-yield nuclear weapons to counter a Warsaw Pact attack with conventional weapons and then, if necessary, escalating to larger nuclear weapons. The final rung in NATO's ladder of escalation is the use of strategic nuclear weapons. Genocide is not abandoned, just delayed. And, of course, now that the Soviet Union has achieved a balance of terror, genocide would be mutual. This is why the doctrine is called nuclear deterrence based on mutual assured destruction, or MAD.

Assured destruction in its pure form is a counter-city doctrine. It is based on the theory that, if the other side knows that most of his cities and industry may be destroyed in retaliation if he attacks you suddenly (pre-emptively), he will not make the attack in the first place. The other side's cities, and civilian population, are the hostages to your nuclear deterrence.

Nuclear deterrence based on assured destruction will only work if it is credible. The other side must believe that the will exists to use nuclear weapons if the situation demands it. The problem is that in almost all situations it would be totally irrational to use nuclear weapons under any circumstances. In Europe, for example, any use of nuclear weapons would be likely to escalate to a point at which, at the very least, the continent would be totally destroyed. In all probability, a nuclear war in Europe would escalate to a strategic nuclear war which would destroy civilization in the whole Northern Hemisphere.

Clearly, no national interest is worth defending at such a cost. Nuclear weapons, in other words, are *militarily* unusable. Their only use is to cancel out the other side's possession of them. US Admiral John Marshall Lee put it this way:

> Nuclear weapons are too powerful, the structure and fabric of nations too fragile, perhaps the ecosystem too vulnerable, to use nuclear weapons operationally. Their entire role is deterring the firing of hostile nuclear weapons. Combat is for conventional weapons (Lee, 1984).

Nevertheless, NATO's military policy still depends on the early and first use of nuclear weapons if war breaks out in Europe. As General Rogers explained when he was NATO's Supreme Commander:

> NATO's current military posture will require us – if attacked conventionally – to escalate fairly quickly to the second response of our strategy, 'deliberate' escalation to

> nuclear weapons. The plain fact is that we have built ourselves a short war; we simply are unable to sustain ourselves for long with manpower, ammunition, and war reserve stocks to replace battlefield losses and expenditures. Therefore, we face the serious risk of having no recourse other than the use of nuclear weapons to defend our soil (Rogers, 1983).

General Rogers has no illusions about the dangers of using nuclear weapons.

> I do not think that a limited nuclear war in Europe is possible. No, if we have to resort to the use of nuclear weapons under current conditions if attacked or if . . . they [the Warsaw Pact] resort to the use of nuclear weapons, in my opinion it cannot be limited to Western Europe. I happen to be one who believes there would be fairly quick escalation to the strategic level. I really think it would escalate and could not be limited.

The General is, of course, aware of the catastrophic consequences of a strategic nuclear war between the USA and the USSR.

The only rational military purpose of tactical nuclear weapons in Europe put forward by General Rogers is that they prevent the Warsaw Pact from concentrating their forces, to avoid giving NATO good targets for its nuclear weapons. If Warsaw Pact forces are unable to concentrate, the General argues, they would be unable to make an effective massive conventional attack. Some Israeli tacticians use similar arguments for the possession and deployment of Israeli nuclear weapons. If they are to serve this purpose, however, the deployment of tactical nuclear weapons must be known to the other side.

If the use of nuclear weapons in Europe would be suicidal and, therefore, irrational, how can they deter war? If nuclear deterrence between two nuclear-armed sides works at all, it depends on making the adversary believe that one would probably actually behave irrationally in a crisis. Thus NATO believes that its threat to use nuclear weapons, even though such use would be suicidal, in some way deters the Soviet Union from starting both a conventional and a nuclear war. The implication is that the Soviets would be so uncertain about whether or not NATO would use nuclear weapons if facing defeat in a conventional battle that they would not risk an attack in the first place. In other words, NATO nuclear deterrence seems to depend on the Soviets believing that NATO political leaders (all of them?) may just be insane enough to use nuclear weapons if facing a Soviet breakthrough with conventional forces. To many, the argument that uncertainty is an adequate deterrent is unreasonably flimsy. Moreover, is the need to convince the other side that one's leaders are most likely to behave irrationally in a crisis a good basis for a credible military posture?

It might be argued, however, that an Israeli threat to use nuclear weapons would be more credible than a NATO or Warsaw Pact threat to do so in Europe. Israel has, after all, demonstrated a great determination to survive against large odds. From past experience, Arab leaders may believe that Israeli political leaders would give permission for the use of nuclear weapons if the population was about to be overwhelmed and feared possible annihilation.

The memory of Hitler's holocaust keeps the fear of a second one in the forefront of the minds of many Israelis. Arab leaders must believe that the Israelis are likely to react to the threat of a second holocaust with all the weapons at their disposal, including nuclear weapons. A second – nuclear – Masada cannot be discounted. In other words, Israel's enemies would be prudent to believe that it has the will to use nuclear weapons, whereas the Warsaw Pact is unlikely to believe that West European countries have the will to use them (and, of course, *vice versa*), given the consequences to them of doing so.

To sum up, the policy of nuclear deterrence by assured destruction rests on four tenets. First, the nuclear forces of the deterrer must be fashioned exclusively for retaliation in response to an attack by the other side's weapons of mass destruction or in response to a threat of annihilation. Second, the nuclear forces – including their command and control systems – must be capable of prompt action. Third, the threat on which the deterrence rests must be the killing of a large fraction of the enemy population and the destruction of much of its economy. Fourth, and most important, the enemy must be aware of the threat in time to deter it from making the actions that will provoke the massive retaliation.

Does nuclear deterrence work?

Apologists for nuclear weapons usually argue that nuclear deterrence has kept the peace in Europe for the last 40 years or so. Some take this argument further and claim that, if nuclear weapons increase security in Europe, they can perform the same function in other regions, such as the Middle East. Avner Cohen disputes this view:

> It is true, of course, that so far circumstances for East-West direct military confrontations have been carefully avoided. But how should we interpret this historical fact? Can we simply say inferentially that this fortunate state of affairs has been achieved *due* to nuclear deterrence? Does it reveal the *intrinsic* stabilizing nature of nuclear deterrence? I don't think so. In fact, one could make the opposite case. It can be argued that the most significant stabilizing factor in East-West relationships during the last forty years – the division of the European continent into two spheres of influence – was achieved prior to and independently of the advent of nuclear deterrence as a significant political factor. Whatever the contribution of nuclear weapons to global security may be, it has been achieved within a political context that has already encouraged stabilization (Cohen, 1986).

Cohen points out that strategists who believe that nuclear deterrence can contribute to stability argue that it can do so only if certain conditions are fulfilled. There must be a bipolar political system; symmetrical strategic relationships; mutual recognition and communication; relative separation between nuclear and conventional deterrence; and invulnerability of second-strike nuclear forces.

> None of these stabilizing elements is currently in place in the Middle East. And considering the history of the region, it is unlikely that they would appear shortly after any move toward nuclear disclosure. In fact, based on the history and geopolitical constants of the region, one can make 'informed' speculations to the contrary. The very act of nuclear disclosure by any state in the Middle East would cause drastic destabilizing effects in the entire region.

Cohen concludes that 'if Israel has actually developed a nuclear option', it did so because its political leaders decided that 'Israel should always be in close proximity to a weapon of last resort' because of 'the conviction that another Jewish holocaust *must* be prevented'. But, he goes on, 'on all conceivable prudential and moral grounds, Israel is obliged to maintain its antinuclear policy and *not* to introduce nuclear weapons politics into the Middle East'. The problem is that if Israel goes on developing more sophisticated nuclear weapons it may not be able to avoid introducing 'nuclear weapons politics' into the region. The very nature of these weapons may make such an introduction inevitable, even if the political leadership would rather not do so.

The commonly held view that the very destructiveness of nuclear weapons precludes the outbreak of nuclear war is false. Even if 'rational' behaviour is assumed, nuclear war is unlikely to occur only if it is believed that neither side can win. If one side perceives a chance of winning, then there is a risk that it will decide to strike while it has the advantage. Moreover, in a serious crisis, the side which perceives that it is at a disadvantage may, if it believes that the use of weapons of mass destruction is likely, attack first and perhaps prematurely, in the hope of reducing the damage it thinks it is almost bound to suffer.

A paradox of the nuclear age is that nuclear deterrence based on assured destruction, if it works at all, does so only with inaccurate nuclear weapons. As more accurate nuclear weapons are deployed the enemy may assume that your nuclear weapons are targeted on his nuclear forces and not on his cities. The cities then cease to be effective hostages. In other words, accurate nuclear weapons weaken and eventually kill nuclear deterrence based on assured destruction.

A relatively small number of nuclear weapons are needed for assured destruction. All that is needed is the number of nuclear weapons required to target the enemy's significant cities. Even in each of the two superpowers, for example, there are at most 200 cities with populations greater than about 100,000 people. Assuming that two nuclear weapons are needed to destroy a large city, about 400 warheads would be more than enough for an adequate minimum nuclear deterrence; more than enough, in fact, to kill roughly 100 million people in each superpower and destroy about half its industrial capability. Nevertheless, each superpower has deployed some 10,000 strategic nuclear warheads – 96 per cent of which are overkill.

Some experts believe that even the superpowers need very few nuclear weapons for adequate nuclear deterrence. Thus, McGeorge Bundy, Special Assistant to the President for National Security Affairs during the Kennedy Administration, wrote:

> In the real world of real political leaders – whether here or in the Soviet Union – a decision that would bring even one hydrogen bomb on one city of one's own country would be recognized in advance as a catastrophic blunder; ten bombs on ten cities would be a disaster beyond history; and a hundred bombs on a hundred cities are unthinkable (quoted in Myrdal, 1976).

If the relations between states, even hostile ones, are being determined rationally, a very small number of nuclear weapons which can be reliably delivered onto their targets are enough for a minimum nuclear deterrent. For the superpowers, this number is certainly much less than a hundred. If the political leaders are irrational, there is no way of calculating how many nuclear weapons would be enough.

Why is it that, if one or ten, or maybe a hundred or so nuclear weapons on target are all that are needed to deter, the USA and the USSR have deployed some 10,000 strategic nuclear weapons? The answer to this question has relevance for understanding nuclear-weapon developments in Israel.

According to Herbert York, a scientific adviser to Presidents Kennedy and Johnson, these numbers are not the result of careful calculations of the need in specific strategic situations. Rather, they are *ex post facto* rationalizations.

> One method for doing so is called 'worst case analysis'. In such an analysis, the analyst starts with the assumption that his forces have just been subjected to a massive preemptive attack. He then makes a calculation in which he makes a series of very favorable assumptions about the attacker's equipment, knowledge and behavior, and a similar series of very unfavorable assumptions about his own forces. Such calculations can result in an arithmetic justification for a very large force indeed, provided that we really believe there is a chance that all the many deviations from the most probable situation will go in one way for them and in the other way for us (York, 1973).

For Israel, a minimum nuclear deterrent, based on assured destruction, against its Arab adversaries could certainly be obtained with a relatively small number of nuclear weapons. Shai Feldman argues that 30 to 40 deliverable nuclear weapons with explosive powers in the 20 to 60 kiloton range would be adequate (Feldman, 1982). Many would argue that the number is much less than 30. There is no rational strategic reason for having more than this number for a policy of nuclear deterrence. A much larger nuclear force is likely to lead to a move away from nuclear deterrence based on assured destruction.

This move is likely if accurate nuclear weapons are deployed. These are, in military jargon, 'counter-force' rather than 'counter-city'. With nuclear weapons capable of destroying even very hardened military targets, nuclear war-fighting based on the destruction of hostile military forces becomes the preferred policy. Accurate nuclear weapons are, in other words, most likely to change the nuclear policy from nuclear deterrence to nuclear war-fighting. This change is likely to occur whether or not the political leadership wants to make the change. It happens because of technological developments.

From nuclear deterrence to nuclear war-fighting

Nuclear war-fighting policies follow from the deployment of accurate strategic and tactical nuclear weapons. If tactical nuclear war-fighting weapons are deployed, they will be integrated into military tactics at relatively low levels of command. The military will then more easily come to believe that if a war occurs nuclear weapons will be used. Nuclear war becomes 'fightable'. And, of course, the military will believe that if they have to fight a war it is winnable.

The belief in the 'fightability and winnability' of nuclear war will make such a war more likely. The deployment of nuclear war-fighting weapons also leads to perceptions that 'limited' and 'protracted' nuclear wars are possible. These also increase the probability that a deliberate nuclear war will occur. And the more sophisticated nuclear-weapon systems become and the more complex nuclear strategies, the greater is the danger that nuclear war will break out by accident, madness or miscalculation.

The most crucial qualitative advance in nuclear weapons is the improvement in the accuracy of delivery. Developments in American nuclear weapons illustrate what improvements are possible.

The accuracy of a nuclear warhead is normally measured by its circular error probability, or CEP, defined as the radius of the circle centred on the target within which a half of a large number of warheads of the same type fired at the target will fall. The Americans have continually improved the guidance system of their intercontinental ballistic missiles so that the CEP has been considerably reduced. For example, at the end of the 1970s, the CEP of the Minuteman warhead was about 400 metres; the new American intercontinental ballistic missile – the MX – has a CEP of about 100 metres.

The MX warhead may eventually be fitted with terminal guidance, in which a laser or radar set in the nose of the warhead scans the ground around the target as the warhead travels towards it through the Earth's atmosphere. The laser or radar locks on to a distinctive feature in the area, such as a tall building or hill, and guides the warhead with great accuracy on to its target. With terminal guidance, MX warheads will have CEPs of 40 metres or so.

The Pershing II tactical ballistic missile – with a range of up to 1,800 kilometres – already has terminal guidance and a CEP of about 40 metres. The Jericho II missile, which uses inertial guidance during the boost phase, could eventually be fitted with terminal guidance and have a CEP comparable with that of the Pershing II. Like the Pershing II, the Jericho II would be a nuclear war-fighting weapon.

Are there any tactical or strategic reasons why Israel would want nuclear war-fighting weapons? Could it be that Israel wants a tactical nuclear arsenal, to be used, for example, against an Arab tank attack? Could its nuclear programme include the production of, for example, neutron warheads for use against massed Arab tank attacks?

At first sight, this may be a plausible explanation of Israel's nuclear activities, including the production of fusion materials. Neutron warheads use fusion to produce high-energy neutrons to irradiate tank crews; the idea is to expose them to such a high dose of neutrons that they are rapidly incapacitated and soon killed. But, in spite of the fact that the Pentagon is producing neutron warheads (W-79 warheads for 8-in. artillery shells), they are ineffective anti-tank weapons. For one thing, tank crews are likely to survive for a few hours even if they have been exposed to enough neutron radiation to kill them eventually. During this time, they are likely to run amok, *kamikaze* fashion, and do far more damage than they would normally. Secondly, if the tanks are spaced about 200 metres apart (a typical separation for tanks in battle), on average only one tank will be caught by a neutron warhead of a reasonable yield. Thirdly, it is relatively easy to protect tank crews against neutron irradiation. A plastic cover, impregnated with boron, will absorb most of the neutrons and protect the crew.

The Israeli military is one of the, if not *the*, most technically competent in the world. It will know the limitations of neutron warheads and would rather spend its money on conventional anti-tank weapons that are much more cost-effective.

Battlefield nuclear weapons are undesirable for another reason. Because they have short ranges most must be deployed near the borders of the country, their main purpose being to attack massed enemy forces as they approach the border. They would, therefore, have to be used very early in a war or they would probably be captured; they are 'use them or lose them' weapons. And, of course, military commanders would rather use them than lose them and run the risk that they may be turned against their own forces. Battlefield nuclear weapons would ensure that an Arab-Israeli war would very rapidly escalate from conventional to nuclear war.

It is, in fact, hard to imagine any rational tactical use for Israeli nuclear weapons. The country is so small that radioactive fall-out would be a major hazard for friendly troops and civilians after any battlefield use of nuclear weapons.

What possible strategic reasons could Israel have for a large nuclear force? One, if not *the*, main aim of Israel's foreign policy is to prevent the emergence of any other nuclear-weapon power in the Middle East. The bombing of the Iraqi reactor in 1981 is an indication of Israel's devotion to this goal. But it is also concerned about the introduction of nuclear weapons into an area larger than the Middle East.

The emergence of Pakistan as a nuclear-weapon power is therefore of considerable concern. Because Pakistan's nuclear-weapon programme has been partly financed by Libya there is a danger that a future Pakistani government may give, or sell, Libya a nuclear weapon. A nuclear-weapon power anywhere in the Islamic world is seen by Israel as a threat to its security.

In particular, Israeli strategists, working, like all strategists, on worst-case analysis, will be considering the possibility that eventually Israel's enemies will acquire reliable and relatively light-weight nuclear warheads and long-range missiles, capable of delivering them accurately. These weapon systems may give their owners the capability to make a sudden pre-emptive attack. Israel's nuclear weapons may then be seen as a counter to such an Arab (or Islamic) nuclear capability. As such, of course, they would be regarded as strategic weapons.

In fact, Israeli strategists may be arguing that, in the medium and long term, Israel may need a strategic nuclear force large enough to target many military targets in the Islamic world. Hostile nuclear forces, nuclear-weapon depots, military command, control and communications centres, and so on, are possible targets in a nuclear war-fighting (or counter-force) strategy.

Once a country starts going down this road, it is possible to find a very large number of targets for nuclear weapons. US military officers, for example, responsible for America's strategic nuclear targeting plans have identified as many as 40,000 targets in the USSR that are reckoned to be suitable for nuclear attack! The rationalization for Israel's extensive nuclear-weapon programme and the development of long-range, accurate surface-to-surface ballistic missiles is very probably based on the need to cover a set of military targets in the Islamic world against the day when other countries in the region acquire nuclear weapons and accurate delivery systems.

And the targets for an Israeli strategic nuclear force may not be limited to Islamic countries. Strategic nuclear weapons may also be targeted on the Soviet Union, particularly the southern part, as a deterrent to possible hostile Soviet moves in some future Arab-Israeli war. The experiences of the 1973 October War, in which the Soviets threatened to prevent Israel's forces from destroying Egypt's army and in which Soviet airborne troops were put on alert, reinforce arguments for such a deterrent.

The threat to use nuclear weapons for massive retaliation in a Middle East war could also be exploited to persuade the Americans to re-supply Israel with conventional weapons. This may, in fact, have been done in the 1973 war to hasten US deliveries of urgently needed munitions. Such a threat is likely to be successful because of fears that the use of Israeli nuclear weapons in a Middle East war would escalate to a strategic nuclear war between the superpowers. This fear is, of course, also likely to encourage diplomatic support from the United States and other countries in, for example, negotiations for a cease-fire on terms favourable to Israel.

To sum up, a number of arguments can be mustered to rationalize Israel's possession of nuclear weapons, however irrational their use in the region may be. The argument most frequently put forward is the need for nuclear weapons to deter the annihilation of Israel in a future conventional war in the Middle East – the last-ditch counter-city scenario. Nuclear weapons are also seen as a way to deter the enemy's use of chemical, biological, or radiological weapons of mass destruction and the use against Israeli cities of very destructive conventional weapons – in particular, those delivered by missiles. Israeli fears of missile attack have recently been enhanced by Saudi Arabia's purchase from China of East Wind 3A missiles with a range of 2,500 kilometres.

The fact that the use of ordinary tactical nuclear weapons on the battlefield could cause considerable death and destruction in Israel itself, may be used to rationalize the development of enhanced-radiation weapons (neutron bombs) on the grounds that these weapons would reduce collateral damage.

Arguments for an Israeli strategic nuclear-weapon force include the need in future to attack military targets throughout the Islamic world, particularly future hostile strategic nuclear targets. It is also argued that Israel needs a strategic nuclear capability against the USSR to deter Soviet involvement on the Arab side in a future war. The threat of the use of nuclear weapons may also be a way of involving the United States on Israel's side in a future Middle East war, particularly to accelerate the delivery of adequate supplies of munitions.

Israeli nuclear weapons are also seen as complementing political goals. By guaranteeing the survival of Israel, they could, it is said, allow it to withdraw its forces from, and give up, the occupied territories, for example. They are also seen as a symbol of total Israeli independence from the United States and other countries. And, by convincing the Arabs that a nuclear-armed Israel cannot be eliminated, nuclear weapons could encourage the Arab states to accept Israel's right to exist and persuade them to negotiate a peaceful settlement in the Middle East.

THE LEGALITY OF ISRAEL'S NUCLEAR WEAPONS

Nuclear deterrence is based on the threat of genocide. Genocide is, of course, illegal. According to the 1950 Nuremberg Principles formulated by the International Law Commission, it is one of the acts contrary to the laws of war that can be punished judicially. The use of nuclear weapons by any country is illegal because they do not distinguish between combatants and non-combatants; they inflict unnecessary suffering on civilians and members of the armed forces alike; they violate the 1907 Hague Convention and the 1925 Geneva Protocol for the Prohibition of the Use in War of Asphyxiating Poisonous or other Gases, and of Bacteriological Methods of Warfare; neutral countries are violated since they will suffer the effects of radioactive fall-out and possibly much more; the legal provisions designed to protect property and undefended cities cannot be kept; and large areas of land are laid waste and the environment is poisoned.

If the use of nuclear weapons is illegal, are their possession and manufacture also illegal? According to Owen Davies, a lawyer specializing in the legality of nuclear weapons:

> There is a strong body of legal opinion which argues that since the use of nuclear weapons is illegal, acts preparatory to their use are similarly illegal. On the one hand, it would appear that the planning and preparation of a nuclear war

> is a violation of the Nuremberg Principles. On the other hand, the production and possession of nuclear weapons *may* have become legal by custom. Neither does the possession of these weapons directly or indirectly infringe upon international agreements (Davies, 1982).

There is, in other words, a gap in international law concerning the general possession of nuclear weapons.

But there are specific reasons for questioning the Israeli possession of nuclear weapons, relating to its use of imported heavy water. The export licence issued by the Norwegian Government in 1959 for the supply of heavy water to Israel was issued on the basis of an agreement between the two governments, and its existence was kept secret until 1979. In that year the SIPRI Yearbook, published by the Stockholm International Peace Research Institute (SIPRI), in a statement about Israel's heavy-water facility, said that it was not known whether the capacity of the plant 'has been sufficient to keep the Dimona reactor in operation after Norway declined further deliveries of heavy water in 1970'. This statement caused sufficient comment in Norway to persuade the Norwegian Foreign Ministry to admit the existence of the heavy-water agreement with Israel, but the agreement itself has not yet been published.

Under the bilateral agreement, the Israeli Government guaranteed that all heavy water sold to Israel by Norway would be used exclusively to promote and develop the peaceful uses of nuclear energy and not for any military purposes. It also guaranteed that none of the heavy water would be handed over to unauthorized persons or transferred outside its jurisdiction without the previous written consent of the Norwegian Government.

Israeli pledged itself to allow Norway to make inspections to ensure that the heavy water was being used in accordance with the conditions spelt out in the agreement. At the time the agreement was signed it was expected that this job would eventually be taken over by the International Atomic Energy Agency (IAEA) when it had established its inspection (safeguards) system. Although the IAEA safeguards system was in operation by the late

1960s, the control of the heavy water has not yet been passed to the Agency. The problem is that, according to the bilateral agreement, Israel as well as Norway must agree to transfer control to the IAEA.

In fact, on only one occasion has an inspection of the heavy water been made on behalf of the Norwegian Government. This was in 1961 by Supreme Court lawyer Jens Hauge. According to Hauge, the heavy water was stored in Israel and he had no reason to believe that it would be used for any purposes other than peaceful ones.

In a statement in the Stortinget (the Norwegian Parliament) on 12 February 1987, Foreign Minister Knut Fryenlund said that the Norwegian Government was 'going to contact Israel with the goal of transferring the security and control functions as mentioned in the agreement to the IAEA'. Clearly, Israel will resist any demands that IAEA inspectors should go to Dimona to inspect the heavy water in the Israeli reactor and to learn the secrets of the nuclear establishment.

If the obligations spelt out in the Israeli-Norwegian agreement are not kept, each of the parties may request that the necessary steps be taken to correct the situation in a reasonable time. Specifically, Norway can ask Israel to explain where the heavy water it has supplied has been since it was first imported. Clearly, the bulk of it is used in the Dimona reactor producing plutonium and tritium for nuclear weapons, and this is a clear violation of the agreement to use it exclusively for peaceful purposes. Such a violation gives Norway the right under the agreement to demand that all the heavy water supplied by Norsk Hydro be returned to Norway. Norway therefore has the power to stop the further production of plutonium and tritium for Israeli nuclear weapons.

Why does Norway not enforce Israel's obligations under the bilateral agreement? In the Stortinget debate of 12 February 1987, the left-wing member Koritzinsky asked the government to inspect the heavy water sold to Israel. In his reply Foreign Minister Fryenlund stated that, for the government, the issue was related to whether or not Israel had nuclear weapons at its disposal; whether the

Norwegian heavy water exported to Israel had been used in accordance with the conditions in the bilateral agreement; whether it was possible today to identify the Norwegian heavy water with a reasonable degree of certainty; and whether Norway was prepared to exercise its right to demand inspections to determine whether or not Israel had used Norwegian heavy water for nuclear-weapon production.

On the first point, the Foreign Minister said that the Norwegian Government did not possess any information giving it reason to believe that 'Israel has nuclear weapons at its disposal'. No foreign intelligence service, he said, had given the Norwegian Government information about Israeli nuclear forces. The Minister went on to explain that the Norwegian Government had asked the Israeli Government for a statement about how the Norwegian heavy water exported to Israel was being used. Not surprisingly, the Israelis replied that it had been used in accordance with the bilateral agreement; it had, they said, been used solely for peaceful purposes.

On the question of the possibility of identifying the Norwegian heavy water, the Foreign Ministry had consulted experts at the Norwegian Institute of Energy Technology who were of the opinion that it would be difficult 'to carry out accurate analyses to ascertain today whether a specific amount of heavy water is identical with the delivered order.' The Minister admitted that the Norwegian Government had made a mistake in not turning over to the IAEA in the 1960s the job of inspecting the heavy water exported to Israel to ensure that it was being used only for peaceful purposes. 'We can only maintain now that, on the Norwegian part, initiatives to transfer the control functions to the IAEA in accordance with the [Norwegian/Israeli] agreement should have been taken.' If they had, the Israelis would, of course, have found it much more difficult, to say the least, to develop a significant nuclear arsenal.

As we saw in Chapter 2, in addition to its imports of heavy water from Norway, Israel obtained some from the United States. According to Professor Gary Milhollin, who has studied in detail the export of American heavy water, this heavy water is still in Israel, and is being inspected (safeguarded) by the IAEA (Milhollin, 1986). The 1955 bilateral agreement obligated Israel to use the heavy water for peaceful purposes only and gave the United States the right to send inspectors to make sure that the obligation was being fulfilled. This right was transferred to the IAEA in 1966 by an American-Israeli-IAEA trilateral agreement. When the 1955 bilateral nuclear co-operation agreement expired in 1977, the IAEA safeguards were extended indefinitely by an exchange of notes and a protocol.

How effective are IAEA safeguards on heavy water? Not very, according to Milhollin.

> IAEA safeguards were virtually non-existent for heavy water until the late 1970s. So, for the period from 1963, when the export was made, or 1966, when safeguards responsibilities were transferred to the IAEA, there was effectively no inspection. Even today, it is not clear how effective IAEA safeguards are for heavy water. They consist of an inspector visiting the country once a year, with plenty of notice, and verifying that the heavy water is in a location subject to safeguards. Since heavy water can be taken out of a reactor quickly, it is entirely possible that Israel could shuffle heavy water around in a way which would satisfy the inspectors (Milhollin, 1987).

Was the American heavy water used in the Dimona reactor? Milhollin gives convincing evidence that it was. His evidence is interesting because it shows how easy it is to muddy the waters and confuse a nuclear safeguards system, if there is an intention to do so. Milhollin studied the reports of the US Department of Energy's Nuclear Materials Management and Safeguards System (NMMSS) – particularly report TJ-7 – and came to the following conclusion.

> In 1977, when the earliest version of the international facility codes was published [by NMMSS], Dimona's code was 'RTSG'. US records show the heavy water going to that facility. The records also show the heavy water at Dimona

(facility 'RTSG') in 1972, when ownership of the heavy water changed to Israel through a bookkeeping transaction. In 1979, however, a new version of the facility codes appeared. It still showed Dimona's code as 'RTSG', but also assigned Dimona a second code, 'RTSL'. Thus, in 1979 Dimona began to have two codes, which contradicts the practice of always keeping the same code for the same facility. In 1982 a third version of the codes appeared; it showed 'RTSG', which had been Dimona's original code, as now designating a facility in Haifa, called 'Technion'. 'RTSL' was shown as designating Dimona. Thus, the current version of the US export record shows the heavy water at Technion, and the earlier versions show it at Dimona. The records do not show whether it was moved. Even if the heavy water is where the records now say it is, it seems to have been at Dimona previously (Milhollin, 1987).

PART II

THE POSITION IN THE REGION AND THE ISLAMIC WORLD

5
The Arab States

Vanunu's revelations leave no reasonable doubt that Israel has a nuclear-weapon capability rivalling in sophistication that of the established non-superpower nuclear-weapon states, like, for example, the United Kingdom – a far greater capability than that necessary for simple deterrence. How will the Arab states react to this information?

Whatever the explanation for Israel's nuclear-weapon programme may be, the Arab states will assume that they are part of a deliberate policy to acquire a pre-emptive nuclear capability. Some of them react by speeding up the development of their own nuclear capability. Others will invest in a chemical-weapon programme, arguing that chemical weapons, being weapons of mass destruction, are, to some extent, equivalent strategically to nuclear weapons.

This is not to suggest that reaction to Israel's nuclear-weapon programme is the sole reason for Arab nuclear- and chemical-weapon activities. Whether or nor Arab countries take the political decision to acquire nuclear weapons, once they have obtained the fissile material, will depend on such factors as prestige and status within the region as well as security.

Of the Arab countries, Iraq, Egypt and Libya have plans for significant nuclear-energy programmes which, if implemented, will enable them to construct nuclear weapons, should they take the political decision to do so. Jordan and Syria have no current plans for such a programme, largely because of a lack of skilled personnel, an insufficient industrial infrastructure, and an unwillingness to foot the large financial bill involved in nuclear-energy programmes. Even Saudi Arabia seems unwilling to divert financial and skilled manpower resources into nuclear energy.

Israel has reason to be relieved that Syria is not investing significantly in nuclear technology. Syria is now its leading Arab opponent and many, if not most, Israelis believe that another war with Syria, although not imminent, is virtually inevitable. But Israel's relief at the absence of Syrian nuclear intentions is greatly tempered by the knowledge that Syria, like Iraq, has acquired the capability to produce chemical weapons.

Arab production of strategically significant arsenals of chemical weapons brings home to Israel that it may not be able to maintain for long its military technological edge over the Arab countries. Moreover, in a future war the Arabs may co-ordinate their activities better. Israel's nuclear programme may therefore include the production of neutron warheads for use against massed Arab tank attacks. This explanation of its nuclear activities is favoured, for example, by Vanunu.

CHEMICAL WEAPONS IN THE MIDDLE EAST

Concern about Israel's performance in future wars is heightened by increasing Arab capabilities in chemical warfare. Authoritative reports of the British Broadcasting Corporation indicate that Iraq is producing 60 tons of mustard gas per month, as well as 4 tons per month of each of the deadly nerve agents Sarin and Tabun. These chemical-warfare agents are produced in a secret plant, some 40 kilometres south of the city of Samarra. This Iraqi military facility, the construction of which began in 1975, is defended by large numbers of Soviet-supplied surface-to-air missiles.

Iraq started using chemical weapons in the Gulf War in 1984. Since then the Iraqis are known to have used mustard gas and the nerve gas Tabun on many occasions, and some Iraqi chemical-weapon attacks have been verified by United Nations experts. In February 1986, for example, about one in ten of the soldiers in a large Iranian force attacking Fao were killed or injured by chemical weapons. Some 2,000 people were reportedly burned with mustard gas on 13 February alone. It is estimated that up to the beginning of 1987 the number of Iranian chemical-warfare casualties exceeded 10,000.

Whereas there is no doubt about the Iraqi use of chemical-warfare agents – a number of Iranian chemical-warfare casualties have been treated in hospitals in Europe – less is known about Syria's chemical-weapon capability. The Syrian chemical-weapon plant is said to be located in a remote desert site north of Damascus. Initially, Syria was reported to have imported chemical weapons from the USSR. But it is now apparently producing chemical agents indigenously, including lethal nerve gases, and the warheads to deliver them, including warheads that can be fitted to SCUD-B and SS-21 surface-to-surface missiles. Syria would, of course, also have the option of using aircraft to deliver bombs filled with chemical-warfare agents.

According to US intelligence sources, Egypt and Libya have some chemical-weapon capability. Egypt is thought to be the first country in the Middle East to have chemical weapons. There were many reports that Egyptian forces used chemical weapons, but not nerve gases, during their intervention in the Yemeni civil war in 1963-7.

The Egyptians may have acquired British chemical weapons left behind when the British forces left Egypt in 1952. German rocket scientists were employed by President Nasser's administration to build missiles for Egypt, and according to some reports, warheads filled with chemical-warfare agents were fabricated for these missiles.

The evidence that Libya has chemical weapons is very scanty. US officials have claimed that Libya has acquired chemical weapons and there have been press reports, attributed to British intelligence sources, of Libyan possession of Soviet-supplied warheads containing nerve gas for its SCUD-B surface-to-surface missiles.

Israel clearly takes Iraqi and Syrian chemical-weapon capabilities seriously. According to a report in the London

Sunday Times of 10 January 1988, the Israeli defence establishment is considering a pre-emptive air strike on Syria's chemical-weapon plant, similar to the 1981 attack on Iraq's Osiraq reactor. Israeli defence experts fear that the Syrian Air Force plans to make a pre-emptive attack on Israeli airfields to render them unusable, thus denying Israel the air superiority it desperately needs in a Middle East war. This fear may provoke Israel to make a pre-emptive attack of its own if it calculates that a crisis in the region may lead to war with Syria.

Be this as it may, Israel is trying to improve its ability to withstand a chemical-weapon attack. The population is, from time to time, drilled on what to do if there is a chemical-weapon attack on population centres, and stocks of gas-masks have been accumulated.

Israel probably has chemical weapons of its own; it would be extremely surprising if it did not. Mordechai Vanunu claimed that chemical weapons were made at Dimona but that he did not know any details about their manufacture. According to a US CIA assessment the Israeli production of mustard and nerve gases began in the 1970s (SIPRI, 1987).

Some Israelis argue that, although the use of chemical weapons by the Arabs, particularly against Israeli cities, may be deterred by Israel's chemical weapons, nuclear weapons are a more credible deterrent. That a chemical-weapon attack in a war will escalate to the use of nuclear weapons and that nuclear weapons deter the use of all other weapons of mass destruction is certainly assumed in NATO. Similar assumptions may determine Israeli policy.

EGYPT'S NUCLEAR HISTORY

Egypt was one of the first Third World countries to initiate a nuclear programme. In 1955 it set up an atomic energy authority; by 1961 it had an operational research reactor supplied by the USSR and a flourishing nuclear research centre at Inshas, 40 kilometres from Cairo. The research reactor has a power output of 2 MWt and uses uranium fuel, enriched to 10 per cent in uranium-235. To date this is the only reactor Egypt is known to possess. It is not capable of producing enough plutonium for a nuclear weapon, and its uranium fuel is not sufficiently enriched in uranium-235 to be used to construct a nuclear weapon.

Egypt has some uranium resources, estimated at about 5,000 tons. A small uranium mine is to be opened, initially producing about 30 tons of uranium, possibly rising to 100 tons, a year (Goldblat, 1985).

For more than twenty years Egypt has considered using nuclear energy as a source both of electricity and of heat to desalinate sea water. The idea is to use the waste heat from nuclear-power reactors (about one-third of the energy from a typical reactor is given off in waste heat) for desalination. The fresh water thus produced would be used for land reclamation – to 'make the deserts bloom'. For a country with so much desert, this application of nuclear power is still a very attractive one.

In the early 1960s, Egypt tried to buy a nuclear-power reactor from France. Doubts about the country's technical ability to operate a power reactor generating a few hundred megawatts of electricity and difficulties about financing the project, as well as arguments about safeguarding the plutonium produced in the reactor, prevented the purchase. Egypt then tried other nuclear suppliers – including West Germany, the USA and China – but met with similar difficulties. It did not lose its interest in nuclear power, however. Plans for an ambitious nuclear-power programme were developed to be put into operation if ways of financing them could be found.

Egyptian politicians link energy consumption with the standard of living. The only way to increase Egyptian per capita income, they say, is to increase per capita energy consumption, which is currently about 20 per cent of the world's average. Furthermore, Egypt's population is increasing rapidly, and, by the year 2000, is likely to stand at about 70 million, as compared to 52 million today. To maintain current standards, therefore, there must be a large increase in energy supply, particularly in electricity generation. But indigenous energy resources are totally inadequate to allow for such a rise and Egyptian planners have concluded that there is no alternative but to import a source, or sources, of electricity.

Studies undertaken by the US Department of Energy and the IAEA in the 1970s recommended that nuclear power should be used to supply much of Egypt's future escalating energy needs. Various Egyptian studies came to the same conclusion. The authorities, therefore, developed a new nuclear-power programme and in 1983 the Minister of Energy and Electricity announced that nuclear power must account for a large fraction of his country's future energy needs. The plan was to build eight nuclear-power reactors by the year 2000, four of them generating 900 megawatts of electricity (MWe) each and the other four generating 1,200 MWe each. The total nuclear generating capacity would account for about 45 per cent of Egypt's predicted electricity demand in the year 2000 (Kats, 1985).

These ambitious plans are based, however, on wrong estimations of the capital costs of nuclear power. According to current estimates, a 900 MWe reactor would cost nearly $5,000 million to build – nearly ten times more than was estimated in the mid-1970s. ($5,000 million is about 8 per cent of Egypt's current gross domestic product.) The decision to go for nuclear power was, however, made on the basis of the early estimates which considerably underestimated the capital costs of doing so. Nevertheless, Egypt is now considering bids for one or two nuclear-power reactors to be built at El Dabaa, near Alexandria. In 1984, however, the Energy and Electricity Minister announced that the plan to build eight reactors by the year 2000 would be delayed by five years. Given the increasing cost of constructing nuclear-power reactors and Egypt's economic situation (external debts already exceed $44,000 million), it is likely that its ambitious nuclear-power programme will be put off again, possibly indefinitely, although one or two nuclear-power reactors may well be built.

Egypt has built up a cadre of highly competent nuclear scientists and technologists. Some 500 of them work at the nuclear research centre at Inshas and at the atomic energy authorities headquarters in Cairo (Spector, 1987). Egyptian nuclear scientists are often appointed to senior international posts, such as with the IAEA. Egypt would have no difficulty in putting together a team to design and construct nuclear weapons, once it had acquired a stock of plutonium and had taken the political decision to acquire nuclear weapons.

EGYPT, ISRAEL AND NUCLEAR DIPLOMACY IN THE MIDDLE EAST

An important event in Egypt's nuclear history was President Nixon's offer, in June 1974, during a visit to the Middle East, to sell nuclear-power reactors to both Egypt and Israel. This offer, which involved reactors with a capacity to generate 600 MWe, surprised everyone but so shocked Israel that it energetically set about preventing the sale of a reactor to Egypt.

Efraim Inbar, a political scientist at the Bar-Ilan University in Ramat Gan, describes the fate of this proposal:

> Israel wanted Nixon's offer to sell nuclear power reactors to Egypt to be part of a package deal that included Israel. This could give Israel a veto on the Egyptian sale. Israel made up its mind to kill the American offer, even at the price of losing its reactors. Nevertheless, Israel continued its negotiations with the Americans over the conditions of supply of the reactors. These pro forma negotiations ended in an agreement signed by Ambassador Dinitz on August 5, 1976. During and after the negotiations, Israel looked for allies within the administration and Congress that shared Israel's apprehensions at supplying nuclear reactors to Egypt. Egypt actually helped Israel's tactics by insisting on overall supervision of nuclear installations in *both* countries. Israel succeeded in resisting US demands for overall supervision, prevented the sale of a reactor to Egypt and also did not appear the guilty party for the failure of the administration's proposal (Inbar, 1986).

The supervision conditions referred to by Inbar involved the extent of the safeguards to be accepted by Egypt and Israel as a condition for buying US nuclear reactors. Up to the 1974 Nixon offer, the US had required safeguards only on any nuclear material or facility that it supplied to a foreign country. But for the sale of reactors to Egypt and Israel, it added the extra condition that both countries should accept international safeguards over any nuclear facilities they might import in the future from any source.

Egypt accepted this condition and, as Inbar explains, suggested that another be added – that both countries accept safeguards over all existing nuclear materials and facilities. In other words, Egypt was suggesting that they accept full-scope safeguards. What Egypt was trying to achieve, of course, was international safeguards on Israel's nuclear facilities and materials at Dimona. Presumably, Egypt calculated that Israel wanted nuclear-power reactors badly enough to agree to such safeguards. If so, it should have known better.

From the worldwide reactions to Nixon's offer and the increasing resistance within the United States during the late 1970s to the export of nuclear facilities to countries outside the Non-Proliferation Treaty, it became obvious to Egyptian leaders that, unless Egypt ratified the NPT, it would find it extremely difficult, if not impossible, to buy nuclear-power reactors from abroad. And, as has been mentioned, Egypt was very keen to buy reactors at that time.

Consequently, Egypt ratified the Treaty on 26 February 1981 and a safeguards agreement with the IAEA has been in force since 30 June 1982. Subsidiary safeguard arrangements for Egypt's research reactor came into force on 31 December 1983.

Why did Egypt decide to ratify the NPT and give up its option to acquire nuclear weapons, after refusing to do so for so long? One of its main objections to the NPT was that it does not include any guarantee by the five recognized nuclear-weapon states to come to the assistance of a non-nuclear-weapon party if it is threatened by nuclear attack. Egypt wanted the nuclear-weapon states, specifically the

USA and the USSR, to give nuclear security guarantees that would pledge them to use, or threaten to use, nuclear weapons to deter the use of such weapons against a non-nuclear weapon state. And, if this deterrence failed, to retaliate against the state that made a nuclear attack.

Egypt may have calculated that, with the signing of the 1979 peace treaty with Israel, the nuclear threat from Israel had been much reduced, or even removed. Its perceived need for nuclear security assurances would then be removed. Be this as it may, the main reason why Egypt joined the NPT in 1981, even though Israel has not done so, was undoubtedly to clear the way to import nuclear-power reactors.

Soon after joining the Treaty, Egypt began discussing the purchase of power reactors from two American firms – Westinghouse and Bechtel – as well as from a consortium of the French firm Framatome and the Italian firm Nira, and from the West German firm Kraftwerk Union. The pace of events demonstrates that President Anwar Sadat's Government still believed that nuclear power was necessary to solve Egypt's urgent energy problems. But, as mentioned above, these negotiations have not yet led to a decision about who should build Egypt's first nuclear-power reactor. If Egyptian enthusiasm for nuclear power has waned, it will be mainly due to economic difficulties, particularly the difficulty of raising capital to finance expensive construction projects. But it is also possible that President Hosni Mubarak's administration is less convinced than its predecessors that nuclear power would be cost-effective for Egypt. If so, it is almost certainly correct.

It can also be argued, as many Israelis do, that one, if not *the*, main reason why Egypt wants nuclear-power reactors is military – to have a source of plutonium if it takes the decision to produce nuclear weapons in the future. Egypt's ratification of the NPT does not mollify the sceptics because of the abrogation clause. Under Article X of the Treaty, Egypt can withdraw from the Treaty by giving 'notice of such withdrawal to all other Parties to the Treaty and to the United Nations Security Council three months in advance'. It can also cancel its safeguards agreement with the IAEA and use any suitable plutonium it has accumulated to make nuclear weapons. By having the non-nuclear components of nuclear weapons already assembled, it could, the sceptics say, have a nuclear-weapon arsenal in a very short time.

In the past, Egypt has tried to prevent Israel acquiring nuclear weapons by seeking nuclear security guarantees from the USA and the USSR; by threatening Israel with pre-emptive military action; and by threatening to produce nuclear weapons itself. In the 1960s, it tried unsuccessfully to persuade the Soviet Union and China to supply it with nuclear weapons. It learnt from bitter experience the rule that existing nuclear-weapon powers are totally committed to keeping nuclear weapons to themselves; the last thing they want to see is an expansion of the nuclear club.

From the late 1970s, Egyptian policy has been to try to persuade Israel to join in the establishment of a nuclear-weapon-free zone in the Middle East (see Chapter 9). This policy makes some sense so long as Israel's nuclear policy is ambiguous enough for Egypt to be able to claim that Israel is not a nuclear threat to Arab countries. In the words of Efraim Inbar:

> It seems that Cairo assesses that Jerusalem has no nuclear bombs in its basement. Such a position could be the result of the Israeli efforts since 1974 to stress its preference for conventional deterrents and to reduce its nuclear ambiguity. The official position in Israel, as well as informal knowledge acquired through the intensive contacts the Egyptians had with the Israelis when negotiating the peace treaty, probably contributed to the Egyptian perception. The Israeli policy of reducing its nuclear ambiguity, accompanied by political agreements, led Egypt to sign the NPT. Obviously, only if Arab countries do not perceive Israel as a nuclear power is there any chance of keeping the Middle East free of nuclear weapons (Inbar, 1987).

Egyptian leaders have always avoided making public statements acknowledging that Israel definitely has a nuclear force. President Nasser threatened to attack Israel if it did produce nuclear weapons. 'We cannot permit Israel to manufacture an atomic bomb. It is inevitable that we would attack the base of aggression', he said in 1960, and repeated the threat several times. President Anwar Sadat has repeatedly stated that Egypt would obtain nuclear weapons if Israel manufactured them. But Egyptian officials maintained that they would only be certain that the Israelis had manufactured nuclear weapons when they tested one. In other words, it seems as if Egypt welcomes Israel's nuclear ambiguity because it allows it to be equally ambiguous about Israel's nuclear capability.

Admittedly Egypt's ratification of the NPT could be taken to mean that it does not believe that Israel has manufactured nuclear weapons or the components for them. From the point of view of Egypt, or for that matter any other Arab country, these two possibilities amount to the same thing. If Israel has made the components for nuclear weapons, it could assemble them very rapidly. The obvious thing would be to assemble the non-nuclear parts of the weapon and keep the nuclear component (i.e., the fissionable material) separately. The task of fitting the nuclear component into the weapon would take a very short time; a sizeable nuclear arsenal could be assembled within an hour. For all intents and purposes, if Israel has made the components of nuclear weapons, it is a nuclear-weapon power.

But it could still claim not to be 'the first to introduce nuclear weapons into the Middle East', relying on the strictly legal definition of a nuclear weapon – i.e. only when the nuclear component is inserted into the non-nuclear assembly. It is hard to believe that Egypt has fallen for Israel's not very subtle deception. A more likely explanation is that since 1979 Egypt no longer perceives Israel as a significant nuclear threat to itself. The Israeli Government's nuclear ambiguity then enables Egypt to justify its attitude to other Arabs.

The details of Israel's nuclear-weapon programme revealed by Mordechai Vanunu, and the credibility given to it by Israel's reaction to the Vanunu affair, are hard for Arab governments to ignore. And so is the virtually unanimous view of the experts on nuclear-weapon proliferation. Once the Arabs generally accept that Israel has a nuclear arsenal far bigger than is necessary for 'last-ditch' nuclear deterrence, Egypt's policy of countering Israel's nuclear-weapon potential by political moves, or by ignoring it, will no longer be credible. Will Egypt then be able to justify any political solution, or will pressures increase within Egypt for nuclear weapons, in spite of the risks involved?

DESTRUCTION OF IRAQ'S OSIRAQ REACTOR

In April 1979, the core of a nuclear research reactor and other reactor parts were blown up in France by a group of saboteurs. The reactor was about to be exported to Iraq and the raid was almost certainly carried out, or organized, by Israeli agents. In June 1980, the head of Iraq's Atomic Energy Authority was assassinated in a Paris hotel.

These events should have alerted Iraq to the lengths to which Israel was prepared to go to prevent the emergence of another nuclear-weapon power in the Middle East. Yet Iraq went ahead with its plans to build a French-supplied reactor at its nuclear research centre at Tuwaitha, about 20 kilometres from the centre of Baghdad.

On 7 June 1981, Israel bombed the Osiraq research reactor just before it was completed. Fourteen Israeli aircraft took part in the raid – eight F-16 Falcons, each carrying two 1,000-kilogram bombs, attacked the reactor, escorted by six F-15 Eagles.

The reactor was completely destroyed, despite the fact that its core was enclosed in a thick concrete container, difficult for bombs to penetrate, and that the F-16s made only one pass at low altitude and very high speed over their target. Under these circumstances it would normally be extremely difficult, to say the least, to destroy such a small, hardened target so completely. This would be so even if precision-guided bombs were used. So far as is known, the American-supplied bombs used were not precision-guided. How did Israel manage to carry out this operation with such surgical precision?

Israel knew that it would have only one chance to destroy the Iraqi reactor. International reaction would ensure that there was no second chance. There are, therefore, good reasons to believe reports that there was also internal sabotage. It is said that explosives had been placed at critical points in the reactor core by Israeli agents – perhaps French technicians working on the reactor and sympathetic to, or paid by, Israel. The explosives could have been actually built into the structure while the reactor was being constructed.

The explosive charges were apparently set off by a radio signal from Israel. There are reports that such a radio signal was picked up at the crucial time by monitors in Jordan. If this is so, it would certainly account for the complete destruction of the reactor, even though the concrete containment vessel was in place.

The Israeli military aircraft had to fly through hostile Saudi Arabian airspace to reach Baghdad. They achieved this even though Airborne Warning and Control (AWACS) aircraft were monitoring the airspace. These aircraft, carrying American operators, very effectively pick up any aircraft, even low-flying ones, flying in the airspace scanned by their radars. The Israelis knew that there was virtually no chance of getting fourteen military aircraft through undetected.

To disguise their aircraft, they used a cunning stratagem. They informed the Saudis in advance that a commercial cargo aircraft would be flying along the same route and at the same time so that the AWACS would feed this information into its computers. The Israeli F-15s and F-16s then flew in a formation that would produce a radar signal similar to that of a large civilian cargo aircraft. Had it not been deceived in this way, the AWACS would have alerted Saudi fighters which would have taken off and investigated the Israeli aircraft. All in all, the complex Israeli operation required a great deal of planning. It was carried out with considerable efficiency and again demonstrated the effectiveness of Israeli intelligence agencies.

There were some costs to Israel. As a consequence of the action against Iraq, the 1982 IAEA General Conference withdrew Israel's credentials. Its participation was not renewed until the 1985 Conference and then only after Israel had committed itself not to attack any nuclear facilities used only for peaceful purposes, whether in the Middle East or elsewhere.

DID IRAQ INTEND TO PRODUCE NUCLEAR WEAPONS?

Iraq's nuclear programme dates back to the early 1960s. Since 1968, a 5 MWt research reactor, supplied by the Soviet Union, has been operating at Tuwaitha. It uses highly-enriched uranium as fuel, but both the reactor and the amount of fuel it uses are too small to be of military significance. A second research reactor, Isis, has been operating at Tuwaitha since 1980. This reactor, supplied by France, has a power output of only 800,000 watts (thermal) and again is of no military significance.

The Osiraq research reactor would have had a power output of about 40 MWt and was scheduled to begin operating at the end of 1981 (Spector, 1987). A reactor of this size would be half as powerful again as Israel's Dimona reactor was when it was first built. France delivered the first batch of enriched-uranium fuel for the Osiraq reactor in mid-1980.

Iraq has a small laboratory-scale reprocessing facility at Tuwaitha, based on three hot cells, bought from Italy. This could be used to reprocess, by remote control, small amounts of irradiated uranium, but it would take many years to produce enough plutonium for one nuclear weapon.

Iraq's eventual intention is to install nuclear-power reactors. To this end, for example, in 1981 it negotiated clandestinely with the Italians for the purchase of a heavy-water nuclear-power reactor. The French also hope that their nuclear co-operation with Iraq will lead to its purchasing French power reactors.

Israel is bound to be nervous about any nuclear activities in Iraq, which it regards as its most dangerous Arab enemy, apart from Syria. Yet Iraq was one of the first countries to ratify the NPT, joining the Treaty in 1969, and while it subscribes to the NPT, all its nuclear facilities are under IAEA safeguards, at least in peacetime. This does not convince Israel, however, that Iraq does not intend to produce nuclear weapons.

The Israeli Government explicitly stated, after its attack on the Osiraq reactor, that it was convinced that Iraq intended to manufacture nuclear weapons to be targeted on Israel. Typical Israeli suspicions were voiced by Professor Yuval Ne'eman, a prominent Israeli nuclear physicist and later Minister of Science, who, according to the *Boston Globe* of 20 July 1980, stated that the only significance of the Osiraq reactor was military.

> There is no other use to which it or the fuel could be put. The French now have in fact supplied the nuclear explosives for bombs which the Iraqis could build and have ready in less than a year.

Israeli nuclear scientists also argue that the Osiraq reactor, although not suited to produce militarily significant amounts of plutonium directly, could have been modified to do so in a matter of weeks by surrounding the reactor core with uranium. The neutron irradiation of the uranium would produce plutonium in the uranium. Israeli scientists calculate that, given a power output of 40 MWt, enough plutonium could, in theory at least, be produced in this way for one or two bombs a year. The allegation that Iraq intended to use the Osiraq reactor in this way is supported by the fact that Iraq has stockpiled large amounts of natural uranium ore from Brazil, Portugal, Niger and Italy. There is also some evidence that it has obtained equipment from Italy for the purification of uranium oxide.

Leonard Spector (1987) offers even more revealing evidence of Iraq's intention to use Osiraq to produce militarily significant amounts of plutonium. He describes how, in 1980, Iraq ordered 11,000 kilograms of uranium-metal fuel pins from the West German firm NUKEM. The pins were of the right size to fit into Osiraq, and their irradiation would have produced about 12 kilograms of plutonium, enough for two nuclear weapons. The deal fell through when NUKEM's subcontractors in the USA and Canada could not obtain export licences for the material.

The Israeli view that Iraq intended to produce plutonium is also backed up by the fact that it imported hot cells which could be used for the small-scale reprocessing of plutonium. The Israelis suspect that Iraq will gain experience, and train personnel, in nuclear and reprocessing technology from its imported facilities and then use the experience and personnel to construct nuclear weapons.

Israel fears have been further enhanced by reports of Iraqi attempts to buy plutonium illegally. According to Leonard Spector:

> In 1982, according to evidence obtained in an ongoing Italian prosecution, Iraq expressed interest in purchasing plutonium from an Italian arms smuggling ring. Senior Iraqi military figures expressed interest in obtaining 34 kilograms of plutonium – enough for several weapons. One member of the ring, interviewed in 1985, claimed that he met first in Baghdad and then in Rome with members of the Iraqi military to discuss the sale; 1982 telexes between members of the smuggling ring contained in the prosecution file are consistent with this claim. The deal fell through when, after a third meeting in Baghdad, the smugglers were unable to produce samples of the nuclear material. Although it is virtually certain that the plutonium offer was a hoax, possibly intended as a come-on to pave the way for sales of conventional arms, the episode suggests that through mid-1982, when negotiations on the matter ended, at least some in the Iraqi government remained interested in nuclear arming.

In mentioning the fuel supplied by France for the research reactors it exported to Iraq, Ne'eman was referring to the fact that the fuel had a uranium-235 concentration of 93 per cent, suitable for use in nuclear weapons. Roger F. Pajak, in his book *Nuclear Proliferation in the Middle East*, comments on the French reply to Israeli criticisms of French exports of weapon-grade uranium to Iraq.

> While the French contract was for a reported 75 kilograms of uranium – theoretically enough for five or six weapons – the French explained that the reactor used only 15 kilograms of the material and indicated that they would ship only that amount of uranium at a time to preclude any unauthorized diversion of the material.

In this way, only 30 kilograms of fuel would be in Iraq, half of it in the Osiraq reactor, and therefore contaminated with radioactivity. The other half was supposed to be slightly irradiated in the Isis reactor to make it unsuitable for use in a nuclear weapon.

The French had many critics, in addition to the Israelis, for their 1976 agreement with Iraq to supply the Osiraq reactor, even though the agreement followed the London Club's guidelines for nuclear exports (see Chapter 8). This criticism was much stronger than that levelled at the French for supplying Israel with the Dimona reactor in the late 1950s.

In replying to the critics, the French were able to point out that IAEA inspectors would visit the reactor to make sure that no nuclear material was diverted to military use. In fact, IAEA inspectors had visited the Osiraq site in early 1981 and inspected the enriched-uranium fuel. Also, it had been agreed that some French technicians would remain at Tuwaitha until 1989 and would probably observe any efforts to divert nuclear materials.

When the Iran-Iraq War got under way, however, France withdrew most of its technicians to get them out of harm's way. But they did not take the enriched uranium supplied to fuel the Osiraq reactor with them; the fuel remains in Iraq. Its fate is uncertain, and is likely to remain so while the war continues.

Roger Pajak quotes the comments of a French source about the unforeseen nuclear problems brought on by the Gulf War:

> We are in a completely new situation that was not foreseen in any international treaties. The problem is raised for international reflection just as sharply as it was in 1974 when India made its [nuclear] explosion.

As the French source said, 'Thought must now be given to finding safeguards against diversion of materials from nuclear facilities in war zones'. The seriousness of the situation was brought home soon after the war began, when Iraq announced that it was not possible to guarantee the safety of IAEA inspectors and, therefore, that IAEA inspections would have to stop.

The Israelis argue, however, that IAEA inspections are of very limited use in Iraq, even in peacetime. Like any country under IAEA safeguards, Iraq can veto specific inspectors, demand that inspectors are of certain nationalities, choose the dates of inspections and postpone them. Thus, since the mid-1970s, the IAEA has been allowed to send only Soviet and Hungarian inspectors to the Iraqi nuclear facilities.

Israel does not believe that ratification of the NPT would be a barrier to Iraq's acquisition of nuclear weapons, if it took the political decision to acquire them. Israel has consistently described the NPT as a weak treaty because the abrogation clause, in Article X, allows any party to withdraw from it after giving three months' notice. It argues that Iraq, for example, may manufacture the nuclear and non-nuclear components for nuclear weapons, withdraw from the Treaty and then quickly assemble its nuclear weapons.

And, in the case of Iraq, it can also argue that it has shown that it is prepared to violate international treaties by its use of chemical weapons, including nerve gases, against Iran. In 1931, Iraq joined the 1925 Geneva Protocol for the prohibition of the use in war of asphyxiating, poisonous or other gases, and of bacteriological methods of warfare. This certainly prohibits Iraq from being the first to use chemical weapons against Iran which is also a Party to the Protocol. The fact that Iraq has violated the Geneva Protocol may mean that it would violate other treaties, such as the NPT, if it chose to do so.

Could Iraq become a nuclear-weapon power?

The June 1981 Israeli bombing was not the first raid on

Tuwaitha. On 30 September 1980, the nuclear centre was attacked by aircraft which dropped several bombs, none of which hit any important buildings. This attack occurred soon after Iraqi forces invded the Khuzistan province of Iran.

The Iran-Iraq War is a drain on Iraq's resources and has considerably set back its nuclear programme; it will presumably be revitalized after the war, with French and/or Soviet assistance. In the meantime, the enriched-uranium reactor fuel supplied by France for the Osiraq reactor could, in theory at least, be used to make a nuclear weapon. Even one nuclear weapon is a major threat to Israel. Fall-out from its explosion at ground level could make a significant fraction of the country uninhabitable and the explosion could kill a large number of Israelis.

It is, of course, possible that Iraq will clandestinely build its own reactor, using natural uranium as fuel, in order to produce plutonium for military purposes. Ironically, if it does, it will be following in Israel's footsteps. And, like Israel, it will have been assisted by French nuclear technology.

The Israeli bombing of the Osiraq reactor and the Gulf War may have provoked Iraq into acquiring such a secret nuclear facility. According to some sources, it will soon start a clandestine military nuclear programme. Indeed in July 1987, the magazine *South* claimed that Iraq was building a secret military nuclear research centre underground at a site near Irbil, under Mount Karochooq, totally protected from air attack. Iraq presumably intends to build the centre without help from Western suppliers or the Soviet Union.

It may get help, however, from countries with which it has had nuclear links for some time – such as Brazil, China and India. And it may get help from Pakistan. Iraq has, in fact, recently agreed to join with Pakistan and Egypt in building small nuclear reactor – according to Leonard Spector, at al-Wadi al-Jadid in Egypt. There is no hint that this nuclear collaboration will have any military implications, but any nuclear collaboration between Islamic states is bound to worry Israel.

I visited the nuclear research centre at Tuwaitha in the late 1970s. The senior nuclear scientists there at that time were highly competent. Typically, an Iraqi nuclear physicist will have spent time abroad in one of the world's best nuclear centres, such as CERN in Switzerland. Admittedly, at that time, there were few such highly educated nuclear scientists with wide experience, but not many are required to design nuclear weapons.

When the Gulf War comes to an end we must expect Iraq's nuclear programme to be revitalized and its nuclear ambitions to be revived and probably strengthened. Countering Israel is not the only reason for Iraq's nuclear ambitions. Iran's nuclear programme under the Shah and its efforts to become the dominant military power in the Gulf have been equally important reasons. And Iraq's ambitions to lead the Arab world must make it aware that a nuclear-weapon force would greatly enhance its status in the region.

France will probably renew its nuclear collaboration with Iraq after the war with Iran is over. Italy may well do the same. France and Italy originally agreed to export nuclear facilities to Iraq in return for guaranteed oil supplies. This motive is as strong as it ever was. It remains to be seen whether next time round the exporters will demand stronger safeguards against military use.

Ultimately, Iraq may import nuclear-power reactors for the generation of electricity. This may seem strange behaviour for a country with large oil reserves. It used to be argued that it made economic sense for oil-exporters to use nuclear power for electricity generation and conserve their oil. But the capital cost of nuclear-power stations has increased so much in recent years that this argument, if it ever was valid, is probably so no longer.

If an oil-producer like Iraq goes in for nuclear-power reactors, this must arouse the suspicion that the main reason is to acquire the capability to produce plutonium and, therefore, a nuclear-weapon option. Even so, if Iraq decides to buy nuclear-power reactors there are likely to be several countries anxious to sell. And the exporters will justify their actions by Iraq's membership of the NPT.

This argument will not impress Israel, however. By using chemical weapons against Iran, Iraq has shown a willingness to deploy and use weapons of mass destruction, a lesson that will not have been lost on Israel. Any future moves by Iraq to acquire nuclear weapons will almost certainly be regarded by Israel as a *casus belli*.

If Iraq does ever decide to build nuclear weapons it has several suitable ways to deliver them, using either surface-to-surface missiles or combat aircraft. It has a number of SCUD-B ground-to-ground missiles, supplied by the Soviet Union. These 300-kilometre range missiles can carry nuclear warheads and could reach targets in Israel. Nuclear-armed SCUDs capable of attacking Israeli cities are obviously a major concern to Israel.

The Iraqi Air Force also operates two squadrons of Soviet-supplied bombers: one with Tu-16 Badger bombers, the other with T-22 Blinder bombers. Both types can carry nuclear weapons. Other Iraqi aircraft that could deliver nuclear weapons are the Soviet-supplied MiG-23BM Flogger; the French supplied Mirage F-1EQ5 (which can deliver Exocet missiles); the Soviet-supplied Su-7 Fitter A; and the Soviet-supplied Su-20 Fitter.

A LIBYAN BOMB?

Libyan plans to import nuclear-power reactors from the USSR, the training abroad of numerous students in nuclear science and technology, the operation of a nuclear research centre at Tajoura, clandestine nuclear assistance to Pakistan, such as supplying uranium obtained by Libya from Niger, Colonel Qadhafi's ambitions for regional status, and nuclear collaboration agreements with Argentina, Brazil, and the Soviet Union are some of the reasons for believing that Libya may – within a decade or so – have the capability to make nuclear weapons. Libya's implacable hatred of Israel and its alleged support for international terrorism make this, for many people, an awesome prospect. Justifiably or not, Libya has the reputation of being the most unstable country in the

Middle East, and Colonel Muammar Qadhafi is reckoned to be the least predictable political leader in the region. According to the conventional wisdom, this is not the sort of country that should be encouraged to obtain nuclear weapons.

Yet a number of countries – including the USSR, Belgium, France, Niger, India, the United States, Pakistan, Argentina, and Brazil – have over the years provided nuclear assistance to Libya. In some cases, the assistance involved the training of Libyan students in nuclear physics, chemistry, and engineering. In other cases, nuclear materials and facilities have been supplied. Some countries have even considered selling Libya a nuclear-power reactor. In 1976, for example, negotiations were held with France for the purchase of a 600-MWe reactor; a preliminary agreement was reached but strong international reaction forced France to cancel the project.

Libya had more luck with the Soviet Union, however. In 1975, the USSR agreed to supply Libya with a research reactor and assistance to establish a nuclear research centre. The centre was built at Tajoura, near Tripoli, and its research reactor began operating in 1981. It has a power output of 10 MWt and uses highly-enriched uranium as fuel. In December 1977, the USSR also agreed to sell Libya two nuclear-power reactors, each capable of generating 440 MWe. The plan, which has not yet been implemented, is to build the reactors on the coast of the Gulf of Sidra, to provide electricity and to desalinate sea water for hotels in tourist centres.

In 1984, Libya tried to negotiate a nuclear collaboration agreement with Belgium. Under it, two Belgian companies which had assisted Libya in the construction of the Tajoura Nuclear Research Centre would help to construct two power reactors. For these services, Libya was prepared to pay a total of $1,000 million. But under pressure from Washington, Belgium in the end withdrew from the deal.

Libya has had a lengthy nuclear collaboration with Argentina and is trying to extend it, but apparently without much success. Eventually, it might have more success in its efforts with Brazil, a major trading partner and a significant supplier of conventional weaponry to Colonel Qadhafi. Libya would find Brazil an attractive nuclear partner because of the latter's growing experience in uranium enrichment and reprocessing technologies.

Libya has obtained a large amount of uranium oxide (U_3O_8, or yellowcake) from Niger. Much of this was re-exported to Pakistan and presumably used in that country's uranium-enrichment programme. Libya has reportedly also given Pakistan some $200 million to help finance its development of nuclear weapons. In return for this collaboration, it is suspected that Pakistan has agreed to provide Libya with the technology to produce the nuclear material for nuclear weapons or even eventually to supply ready-made nuclear weapons (Spector, 1987).

Past Libyan efforts to buy nuclear weapons 'off the peg' from anyone willing to sell have failed. In 1970, for example, Colonel Qadhafi asked the Chinese political leaders to sell him a nuclear weapon but they refused (US Government Printing Office, 1981).

Libya has a number of Soviet-supplied SCUD-B missiles which could deliver nuclear weapons. In addition, the Libyan Air Force operates a number of Soviet-supplied Tu-22 Blinder bombers and other types of aircraft that could carry nuclear weapons. These are the French-supplied Mirage 5D/DE and Mirage F-1AD, and the Soviet-supplied MiG-23BM Flogger F and the Su-20/22 Fitter.

Libya ratified the Nuclear Non-Proliferation Treaty in 1975. But Colonel Qadhafi's statements about his desire for nuclear weapons and his clandestine efforts to acquire them cast doubt on the strength of Libyan commitment to non-proliferation.

6
An Islamic Bomb

Although Pakistan and Iran are outside the Middle East region, and in general have different geopolitical and strategic considerations from those of Israel and the Arab countries, any moves by them towards a nuclear-weapon capability have considerable consequences for Israel. For this reason, this chapter will consider the nuclear activities and ambitions of the two countries.

Pakistan's former Prime Minister Ali Bhutto wrote as early as 1969:

> Our plans should include the nuclear deterrent. . . . If Pakistan restricts or suspends her nuclear programme, it would not only enable India to blackmail Pakistan with her nuclear advantage, but would impose a crippling limitation on the development of Pakistan's science and technology. Our problem, in its essence, is how to obtain such a weapon in time . . . (Bhutto, 1969).

In 1972, following Pakistan's humiliating defeat by India in Bangladesh, Bhutto was convinced that India was producing a nuclear weapon, a fear confirmed in May 1974 when India set off a nuclear explosive. Consequently, he called a secret meeting of 50 or so Pakistani scientists and told them that Pakistan should start a nuclear-weapon programme. Like Israel, Pakistan is an example of how it is possible for a non-industrialized country to acquire a nuclear-weapon capability illicitly, despite international controls specifically designed to prevent the spread of nuclear weapons.

Although the nuclear programmes of Pakistan and Iran were stimulated by factors other than a response to Israel's nuclear capability, Israeli nuclear strategy must take account of them. The prospect of an 'Islamic bomb' greatly complicates Israel's foreign and security policies. Israel must now worry about the introduction of nuclear weapons into the entire Islamic world rather than just the Middle East region.

The governments of Pakistan and Iran have both been involved in some anti-Israel posturing, but in reality this may not mean very much. The Khomeini regime – despite its clandestine arms deals with Israel – claims to be anti-Zionist and threatens to 'liberate' Jerusalem. Pakistan rhetorically uses the Islamic bomb threat against Israel, and has accepted aid for its nuclear programme from both Saudi Arabia and Libya. Statements like that made by President Zia ul-Haq in March 1986, and quoted by Leonard Spector (Spector, 1987), enhance Israeli fears:

> Why must they call Pakistan's bomb, supposing we have it, an Islamic bomb? You can see the mentality. They are fearful that if an Islamic country such as Pakistan acquires this technology they will spread it. In fact if the Islamic world possessed this technology, it means that 900 million Muslims possess advanced technology. Here comes the aggressive campaign against Pakistan and the aggressive talk about the Pakistani nuclear bomb. It is our right to obtain the technology. And when we acquire the technology, the entire Islamic world will possess it with us.

In Spector's judgement the President's comments refer to a 'symbolic sharing of uranium enrichment technology, rather than to the actual sharing of this technology or of nuclear weapons'. Be this as it may, Israeli strategists are bound to make the worst-cast analysis and worry that, under some circumstances, a future Pakistani administration may make nuclear weapons available to Arab countries.

Whereas Israel's nuclear-weapon programme is based mainly on plutonium-239 as the fissionable material, Pakistan's programme is based on enriched uranium.

Pakistan may have decided to go the uranium route because it obtained access to secret information from the Netherlands that enabled it to build the necessary gas centrifuges. It may be that the Pakistani authorities calculated that, given this information, they could build a plant independently and avoid international safeguards. It may also have been more risky at that time – the mid-1970s – to build a clandestine reactor to produce plutonium and a secret reprocessing facility to remove the plutonium from the spent reactor fuel elements.

There is no doubt that Pakistan has the knowledge and experience to go the plutonium route, but it must have decided that it was politically safer not to do so. For one thing, it would have been very hard to prove that, if plutonium was used to produce a nuclear explosion, it had not come from the power reactor supplied to Pakistan by Canada. Any suspicion that this had happened would have provoked Canada to protest vigorously. The fact that India had used plutonium produced in a reactor built with Canadian assistance for its 1974 nuclear explosion still rankles with Canada. Canada demonstrated its disapproval by immediately withdrawing its aid to India. Pakistan would obviously prefer to avoid this sort of trouble. By building an enriched-uranium plant indigenously Pakistan can argue that it has not breached any agreements with its nuclear suppliers.

Nevertheless, there have been rumours that it may clandestinely construct a reactor specifically for the production of weapon-grade plutonium for nuclear weapons. Suspicions of such intentions have been enhanced by the government's announcement in 1984 that it was producing high-grade graphite and its reported unsuccessful attempts to buy a few hundred tons of pure graphite in the UK, France, West Germany and the USA (Spector, 1987). With graphite, Pakistan could construct a simple reactor, fuelled with natural uranium, for the production of military plutonium.

THE GUN TECHNIQUE OR IMPLOSION?

As has been seen, a nuclear weapon using enriched uranium as the material which undergoes fission can be based on the 'gun' design which involves the bringing together of two masses of fissionable material, each of which is less than the critical mass but when combined forms a supercritical mass. For example, a subcritical mass of enriched uranium is propelled, using a conventional explosive, down a 'gun barrel' (a strong metal tube) into a second subcritical mass at the bottom of the tube. At the moment the two masses meet, forming the supercritical mass, a neutron source is assembled to ensure a burst of neutrons to start the chain reaction. In the bomb dropped on Hiroshima, the second mass was encased in a steel tamper lined with tungsten carbide. Nowadays, the gun device is rarely used, although the W33 US 8-in. artillery shell still uses it.

One advantage with the gun assembly is that it is fairly simple and a reasonably competent designer would ensure that it would work as predicted. In fact, as we have seen, the designers of the Hiroshima bomb were so confident that it would work that, even in those days, no test was made of the design.

The disadvantage of the gun technique, compared with the implosion technique, is that it requires a relatively large amount of fissionable material; the Hiroshima bomb used 60 kilograms. For a given amount of enriched uranium, implosion produces a much larger explosive yield than the gun assembly (see Appendix for more detail). Thus, fission nuclear weapons in a modern nuclear arsenal, like that of the United States, even if they use only enriched uranium as the fissionable material, use the implosion, rather than the gun, technique. It is believed that Pakistan is using an implosion design in its efforts to make nuclear weapons.

PAKISTAN'S NUCLEAR PROGRAMME

Pakistan has had a significant nuclear programme since the early 1960s. Over the years, a group of very competent nuclear scientists and technologists have been trained, among them the Nobel Prize winner Abdul Salam.

Pakistan has some uranium resources – estimated, somewhat uncertainly, at about 20,000 tons. There is a uranium mine at Bagalchur, near Multan, but it is not publicly known how much uranium ore is mined. It presumably produces enough to fuel the KANUPP nuclear-power reactor near Karachi and for the Kahuta uranium-enrichment plant. There are two uranium mills in operation in Lahore.

A research reactor, supplied by the USA, started operations in 1965 at Rawalpindi and has a power output of 5 MWt. The reactor uses about 5 kilograms of highly enriched (93 per cent) uranium as fuel, also supplied by the USA.

The KANUPP nuclear-power reactor supplied by Canadian General Electric, and moderated and cooled by heavy water, went into operation in 1972. This reactor is capable of generating 125 MWe. Canada supplied the natural uranium fuel for the reactor until 1976 but then stopped doing so because of Pakistan's refusal to ratify the Non-Proliferation Treaty. Since 1980, Pakistan has produced its own fuel for the reactor.

It has also obtained uranium from other sources. In 1977, a safeguards agreement was signed with the IAEA for the import of yellowcake from Niger. It has also obtained supplies of uranium clandestinely. *The Washington Star*, for example reported on 25 November 1979 that Pakistan acquired some 100 tons of yellowcake from Libya. This uranium, which was not subject to IAEA safeguards, was first bought from Niger by Libya and then sent on to Pakistan.

With components supplied by the West German firm CES Kalthof GmbH of Freiburg, Pakistan has built a plant to convert yellowcake to the gas uranium hexafluoride

(UF_6) – the form in which uranium is enriched in gas centrifuges. The plant went into operation in 1982 at Multan and produces about 200 tons of hexafluroide a year (Goldblat, 1985).

A fuel fabrication plant has also been built, using Canadian plans, to supply fuel rods for the KANUPP reactor. The reactor needs about 15 tons of natural uranium fuel a year. Presumably, the plant is capable of supplying this amount.

The initial supplies of heavy water for the KANUPP reactor came from the United States and Canada, but Pakistan has since built two heavy water plants of its own. One, at Karachi, is a small plant, possibly supplied by Canada, which came into operation in 1976 to serve KANUPP. The other, built at Multan, went into operation in 1980 and is said to produce some 13 tons of heavy water a year and to have been supplied by the Belgian firm Belgonucléaire.

According to current plans, another nuclear-power reactor, generating 900 MWe, will be operating by the mid-1990s, but the supplier of this reactor has yet to be decided. It is planned that four more nuclear-power reactors will be operating by about the year 2000, with a total power output of 2,800 MWe. If all these reactors are constructed they could generate about 30 per cent of today's total electricity production in Pakistan.

If in addition to these known reactors, Pakistan decides to construct a military plutonium-production reactor clandestinely, it will have considerable experience in the technology for reprocessing any plutonium produced. For many years now, a hot cell laboratory, capable on a small scale of removing the plutonium from spent reactor fuel elements, has been operating at the Pakistan Institute for Nuclear Science and Technology (PINSTECH) at Rawalpindi near Islamabad, the biggest of several nuclear research centres in Pakistan. The laboratory was built with assistance from Belgonucléaire with a capability of handling a ton or two of reactor fuel elements a year. In 1981, this laboratory plant may have been upgraded, with Chinese help, to reprocess 10 to 20 tons of spent reactor fuel and produce 10 to 20 kilograms of plutonium a year.

The New Labs reprocessing plant at Rawalpindi is said to handle about the same amount of spent reactor fuel – 10 to 20 tons a year – and produce the same amount of plutonium – 10 to 20 kilograms a year. It was cold tested in 1982 and may have been operating since 1984, although this is uncertain. The plant was supplied by Saint-Gobain Techniques of France and Belgonucléaire of Belgium.

The reprocessing facilities of Rawalpindi, although either laboratory scale or pilot plants, are probably capable of producing enough plutonium for a few nuclear weapons a year. The Chashma plant is a much bigger reprocessing plant. In 1976, France signed an agreement with Pakistan under which the French firm Saint-Gobain Techniques Nouvelles would build the plant, which was to be able to handle 100 tons of spent fuel and produce about 100 kilograms of plutonium a year (Spector, 1987; SIPRI, 1985). After considerable pressure from the United States, as part of its policy of discouraging the proliferation of nuclear weapons, France reluctantly withdrew from the project in 1978. But its construction may be continuing. Indeed, it is reported that Pakistan has nearly completed the construction of the plant with the clandestine help of the French firm and the help of Italian, Swiss and Turkish companies (Spector, 1987).

If Pakistan decides in the future to go for a large nuclear-weapon force, it may use plutonium-239 from a specially built plutonium-production reactor as well as uranium-235. If it does, it could easily build a clandestine reprocessing plant to acquire the plutonium.

The uranium-enrichment plant at Kahuta

Pakistan has also received substantial help from a number of foreign companies to build its top-secret uranium-enrichment plant at Kahuta, about 30 kilometres south-east of Islamabad. Leonard Spector lists the following firms as supplying components for the Kahuta plant: Vakuum Apparat Technik and CORA Engineering of Switzerland;

Emerson Electric of the UK; Van Doorne Transmissie of the Netherlands; and Leybold Heraeus and Aluminium Walzwerke of West Germany. Sundry items were also obtained from the United States and Canada, and design information was clandestinely obtained from URENCO, the enrichment plant at Almelo in the Netherlands.

The Kahuta plant contains a large number of gas centrifuges. No important details about the plant have been publicly released and it is very heavily guarded. Western journalists who have tried to get near it have been arrested by security guards and on some occasions beaten up. There have, however, been a number of educated speculations about the plant's capacity.

One thing that is reasonably certain is that the centrifuges are similar to those used at the URENCO plant. The designs of the URENCO centrifuges were almost certainly acquired by the Pakistani engineer Abdul Qadeer Khan, when he worked between 1972 and 1975 for a Dutch engineering company which was closely associated with the URENCO plant. On leaving the Dutch firm Khan went back to Pakistan and started the Pakistani uranium-enrichment programme. He took home with him a great deal of secret technical information about gas centrifuges suitable for uranium enrichment and lists of the equipment needed for them. We can confidently assume that the Kahuta centrifuges are based on knowledge acquired by Khan in the Netherlands. This was the conclusion of the investigation into Dr Khan's activities in the Netherlands made by the Dutch Ministry of Foreign Affairs. Its report concluded that Khan took 'essential gas centrifuge know-how' to Pakistan.

But Khan learnt about more than one type of ultra-centrifuge. We do not know which types are installed at Kahuta, although there is evidence that two types are used. It is reported that Kahuta, which began operating in early 1984, has two centrifuge 'halls', one containing centrifuges made from aluminium, the other centrifuges made from specially-toughened Maraging steel. That some of the centrifuges are made of steel is supported by an article in *Der Stern* of 30 April 1986. An ex-worker at Kahuta told the magazine that centrifuges frequently blew up because of 'imbalances in the steel'.

The amount of weapon-grade uranium being produced

The total output of a centrifuge plant will, of course, depend on the number of centrifuges in the plant and their separative capacity. There are various estimates of the number of centrifuges operated at Kahuta. According to the Stockholm International Peace Research Institute, Kahuta currently has 1,000 centrifuges and plans to have between 2,000 and 3,000 (Goldblat, 1985). This estimate, made in 1985, was repeated on 9 August 1986 by the Pakistani newspaper *Muslim*, which described the estimate as a 'rumour'. US Senator Alan Cranston made a similar estimate in the *Congressional Record* of 21 June 1984.

Other estimates are very much higher. Leonard Spector quotes the *Economist's Foreign Report* of 1 May 1986 which claims that the number is as high as 14,000 (Spector, 1987). This number was also published in *Der Stern* of 30 April 1986. According to David Albright, a scientist working at the Washington-based Federation of American Scientists, in an article in the June 1987 *Bulletin of the Atomic Scientists*, 'US government sources confirmed that 14,000 centrifuges is a more accurate count than the lower estimates' (Albright, 1987b). But the sources did not say how many of these centrifuges are actually in operation at Kahuta.

Estimates of the amount of weapons-grade enriched uranium produced by Pakistan also vary considerably. SIPRI (1985) quotes an estimate of 45 kilograms a year, the figure also given by Senator Cranston. But there is no indication of when this rate of production, enough for about two or three nuclear weapons a year, will be reached. (A very competent modern nuclear-weapon designer could design an implosion-type nuclear weapon with about 15 kilograms of uranium which contains more than 90 per cent of uranium-235; a weapon needing 25 kilograms would today be regarded as a crude device.)

Foreign Report and *Stern* estimate the amount produced

at Kahuta to be 10 kilograms a year from 1987. But although these reports give more definite figures they do not explain how their estimates are made. The various estimates are not consistent with each other and vary in relation to the number of centrifuges estimated in the reports.

Albright bases his calculations on the separative capacity of the centrifuges at Kahuta. He estimates this capacity by considering the performance of centrifuges at the URENCO plant at Almelo in the mid-1970s, when Khan was associated with it.

The performance of a gas centrifuge used to enrich uranium is measured in units called separative work units (SWUs). These units are complicated because the amount of work needed to enrich uranium to a given concentration of uranium-235 depends on the concentration of uranium-235 in the uranium hexafluoride fed into the centrifuge and in the uranium in the waste gas at the end of the centrifuging operation. For example, to produce one kilogram of uranium enriched to 94 per cent of uranium-235 (i.e., weapons-grade uranium) requires 238 SWUs, assuming that the input into the centrifuge is natural uranium and that the concentration of uranium-235 in the waste is 0.2 per cent.

Because each gas centrifuge can contain only a small volume of uranium hexafluoride gas, a large number of centrifuges are needed to produce the kilogram quantities of highly-enriched uranium needed for a nuclear weapon. The centrifuges are connected in a cascade so that the uranium hexafluoride is increasingly enriched in uranium-235 as it passes from one centrifuge to the next through the cascade.

Performance details of URENCO's centrifuges are closely guarded secrets. But some information about the performance of a German-designed centrifuge known to have been used in the URENCO plant in the mid-1970s has been published. The capacity of this centrifuge is 5 SWUs per year (Boskma, 1974).

Khan would most likely have had access to the design of centrifuges more advanced than the early German type and a capacity greater than 5 SWUs a year. On the other hand, the Pakistanis may have decided to go for a more proven design with a capacity less than 5 SWUs. In the absence of any evidence about the precise type of centrifuge chosen, Albright bases his estimates on a capacity of 5 SWUs a year for each centrifuge. He points out that the Kahuta centrifuges are 'supercritical machines' (Albright, 1987b). This means that they

> have rotors operating above critical rotational frequencies or resonances. That is, certain rotor speeds create resonant vibrations, and if the rotor remains at or near resonance for any length of time, the vibration can increase until the rotor or its bearings are wrecked. These centrifuges, therefore, must be operated at speeds away from resonant frequencies and must be able to accelerate rapidly through resonant speeds.

Albright accepts the estimate that Pakistan has built 14,000 centrifuges but, given the problems in operating such ultra-speed centrifuges, he believes that the number actually currently operating is closer to 1,000. The story told to *Der Stern* by the ex-worker at Kahutu, that many centrifuges crack up because they cannot withstand the huge centrifugal forces produced in them, supports Albright's conclusions.

If Kahuta has 1,000 centrifuges working, on average, each with a separative capacity of 5 SWUs per year for a total of 5,000 SWUs per year, the plant could produce annually a total of 21 kilograms of weapon-grade uranium (enriched to 94 per cent of uranium-235). This would be more or less enough for one nuclear weapon a year.

As the scientists and engineers gain experience in operating ultra-centrifuges, they will be able to keep an increasing number in continuous operation. If they have already built 14,000 centrifuges, they will eventually move towards a separative capacity of at least 70,000 SWUs per year. This will produce about 300 kilograms of weapon-grade uranium, which should be enough for at least 15 nuclear weapons a year.

According to Spector (1987), Pakistan has been operating an experimental-scale ultra-centrifuge plant at Sihala since at least 1984. Kahuta has also been operating, at least partially, since 1984. Pakistan therefore has a supply of low-enriched uranium, which, Albright points out, could be used as the input material for a special ultra-centrifuge plant producing weapon-grade uranium.

> Because of the increased failure rates while stopping and starting centrifuges, producing weapon-grade uranium from low-enriched uranium feed might be accomplished more effectively in two separate cascades – one that enriches natural uranium to a few per cent and another that enriches this product to weapon grade. If Pakistan has dedicated only 700 centrifuges to producing 5 percent enriched uranium from the beginning of 1984 through 1986, it could have produced about 1,200 kilograms of 5 per cent enriched product. If it now uses 400 other centrifuges to enrich the 5 per cent product to weapon-grade uranium, Pakistan could produce about 50 kilograms of weapon-grade uranium a year, assuming the waste contained a 1 per cent concentration of uranium 235. This is enough weapon-grade material for two nuclear explosives (Albright, 1987b).

Albright's suggestion is interesting in the light of a report from Simon Henderson in the London-based *Financial Times* of 11 December 1987. According to Henderson, Pakistan is building another uranium-enrichment plant at Golra, some 10 kilometres west of Islamabad. The Golra plant, whose construction is being monitored by US reconnaissance satellites, is expected to use centrifuges made from Maraging steel. Although a centrifuge hall has been constructed, the centrifuges have yet to be installed.

Incidentally, Dr Khan has been associated with Golra, which is a military base, for some years through the Precision Engineering Division, which is one of his responsibilities. A facility at Golra is said to be currently producing precision (non-nuclear) components for nuclear weapons.

It is possible that if a uranium-enrichment plant is completed at Golra, it will produce weapon-grade uranium using as feed material uranium enriched to, say, 5 per cent. Kahuta could then be limited to 5 per cent enrichment, and some of its output could then be sent to Golra, while the rest could be used to produce fuel elements for future nuclear-power reactors. It may then be possible to open Kahuta to IAEA inspection. It should be noted that the work needed to increase the concentration of uranium-235 in uranium from its natural value of 0.7 per cent to 5 per cent is about 70 per cent of that needed to go from natural uranium to over 90 per cent enrichment, i.e., to weapon-grade uranium.

PAKISTAN'S NUCLEAR AMBIGUITY

The Pakistani Government has been deliberately ambiguous about its nuclear-weapon programme throughout. Uranium enriched to 5 per cent of uranium-235 can be used as fuel in nuclear-power reactors. And the production of power-reactor fuel rather than nuclear weapons is the official reason given for the uranium-enrichment programme. This claim is made despite the fact that Pakistan's only power reactor uses natural uranium as fuel. The Pakistani authorities get around this difficulty by arguing that Pakistan will need low-enriched uranium to fuel the 900-MWe power reactor it intends to import and operate in the mid-1990s! This claim would be more credible if Pakistan had actually ordered the reactor or even chosen the supplier. Given that it takes well over ten years from the time a nuclear-power reactor is ordered to the time it goes into operation, it is increasingly unlikely that Pakistan will fulfil its ambition of increasing its nuclear electricity-generating capacity before the year 2000.

Pakistan's ambiguity about its nuclear programme took a bizarre turn when the London newspaper *The Observer* published, on 1 March 1987, an interview given by Dr Khan to Indian journalist Kuldip Nayar. In it, Khan removes all doubt that Pakistan can produce a nuclear weapon: 'What the CIA has been saying all the time about possessing the bomb is correct'. He explained that it was difficult for Pakistan to get hold of the technology, 'but we purchased whatever we wanted to before the Western countries got wind of it [Pakistan's nuclear-weapon programme]'. When asked about 'peaceful' uses, Khan gave an honest answer:

> The word 'peaceful' associated with a nuclear programme is humbug. There is no peaceful bomb. . . . Once you know how to make reactors, how to produce plutonium – all that Pakistan has mastered as well – it becomes rather easy to produce nuclear weapons.

On 2 March 1987, the day after *The Observer* interview appeared, Khan claimed, in an article in the *International Herald Tribune*, that some of his remarks were taken out of context to

> mislead the world into believing that Pakistan possesses a nuclear weapon and that we have enriched uranium to 90 per cent or more. As I have so often publicly stated, Pakistan's enrichment research is solely aimed at the development of fuel-grade plutonium for our future power reactors. The government of Pakistan has made it abundantly clear that it has no desire to produce nuclear weapons.

Khan was repeating assurances given earlier by Prime Minister Mohammad Khan Junejo to President Reagan that the Kahuta plant would not produce uranium enriched to more than 5 per cent of uranium-235. The President, however, is reported to have had CIA information that Pakistan has produced uranium enriched to 93.5 per cent.

Even so, President Mohammed Zia ul-Haq, in an interview in *Time* magazine of 30 March 1987, said: 'Pakistan has the capability of building the Bomb. You can write today that Pakistan can build a bomb whenever it wishes'. He hastened to add, however, that it had no plans to construct nuclear weapons.

Pakistan's ambiguity about its nuclear-weapon programme is, like Israel's ambiguity, intended, partly at least, to confuse the US Congress. Under the 1976 Symington Amendment to the Foreign Assistance Act, economic aid and military grants or credit (excluding cash sales of military equipment) must be cut off from any country that imports uranium-enrichment technology, materials or equipment without accepting IAEA safeguards on all its nuclear facilities and agreeing that any resulting enrichment plant will be placed under multinational management.

US assistance to Pakistan was cut off under the Symington Amendment in 1979 when it became known that Pakistan had illicitly obtained centrifuge technology from the Netherlands. In 1981, however, in a policy of encouraging Pakistan's support of the Afghan rebels after the Soviet invasion of Afghanistan, Congress passed a special law waiving the Symington Amendment in respect of Pakistan and authorizing a six-year, $3,200 million programme of economic and military aid. Four years later the Cranston Amendment was passed. This requires that, in order to supply economic aid or to sell military equipment to Pakistan under any legislation, the US President must certify to Congress that Pakistan does not possess a nuclear explosive device and that the assistance in question will reduce significantly the risk that it will acquire a nuclear explosive device (Van Doren, 1987). This assurance was given by the President in fiscal years 1986 and 1987.

On 30 September 1987, the waiver authority expired and the US aid programme to Pakistan came to an end. New economic and military aid cannot be granted without new legislation. The Reagan Administration had by then requested a six-year, $4,020 million programme of aid to Pakistan.

Pakistan will remain in violation of the Symington Amendment until it accepts full IAEA safeguards and multinational management of its enrichment plant(s), or unless the US President certifies to Congress that he has received reliable assurances that it will not develop nuclear weapons. The President must also give Congress the assurances required under the Cranston Amendment before new aid can be granted. This Reagan did early in 1988 and Congress agreed to provide two and a half years of additional foreign aid to Pakistan without imposing new conditions (Carnegie, 1988).

HOW PAKISTAN ACQUIRED FOREIGN TECHNOLOGY

When Dr Khan told *The Observer* that Pakistan had 'purchased' whatever technology and materials it needed for the Kahuta enrichment plant from Western countries before they realized the purpose of the materials, he did not add that much more was obtained illicitly. His own acquisition of the design of ultra-centrifuges and the materials needed to construct them was, of course, a form of industrial espionage. The success of Pakistan's efforts shows how easy it is to acquire foreign nuclear know-how, equipment and materials clandestinely, even if governments try to prevent this acquisition.

Since 1979, Pakistan has made great efforts to obtain materials and technology for its uranium-enrichment and other nuclear-weapon activities. Leonard Spector, who closely monitors these Pakistani efforts, discovered that it has involved the use of

> dummy corporations and transshipments through third countries [which] has by now been well-documented in the press and officially recognized by West Germany, Canada, Britain, the Netherlands, and the United States either in the course of prosecutions or in published government reports.

Perhaps the most egregious case involved the smuggling from West Germany between 1977 and 1980 of an entire plant for converting uranium powder into uranium hexafluoride, the easily gasified material that is the feed-stock for the Kahuta enrichment facility. The plant, located in the town of Dera Ghazi Khan, is continuing to operate, producing material that is essential for the Pakistani nuclear-weapons effort. (Spector, 1987).

Another well-publicized example of the smuggling of nuclear components to Pakistan is the Vaid case. In October 1983, Nazir Ahmed Vaid, a Pakistani from Lahore, visited the American firm EG & G, in Salem, Massachusetts. EG & G is the only company in the western world that commercially manufactures krytons – very high-speed, extremely reliable electrical switches that deliver an extremely precise pulse of electrical current in a very short time (about a millionth of a second).

Krytons are used for a number of highly specialized purposes – in high-energy lasers, for example. But their most important function is as part of the triggering mechanisms for nuclear weapons. They are easy to buy if one is in the United States; they cost about $90 each. But, because of their use in nuclear weapons, overseas sales are strictly licensed and monitored by the US Government.

Vaid bought 50 krytons. He told EG & G that he was commissioned to buy them by the University of Islamabad. But, as the EG & G employees knew, there is no conceivable reason why the University would want 50 krytons. They would be required, however, for a nuclear-weapon programme. The EG & G employees promptly informed the FBI which asked them to sell Vaid the krytons.

When Vaid obtained them he tried to air freight them to Pakistan, and it was when he tried to get them through US customs described as office supplies including 50 'bulbs/switches' that he broke the law. He was arrested, tried, served three months in jail, and was discreetly deported. The judge thought that he was an ordinary businessman who had been rather careless about US export regulations – hence the light sentence. But Seymour Hersh, a *New York Times* journalist, discovered cables showing that Vaid, far from being an ordinary businessman, was a Pakistani agent in direct contact with senior officials of the Pakistan Atomic Energy Commission about the purchase of the krytons. It turned out that the krytons were actually ordered by a director of supply and procurement for the Commission.

A second case of nuclear smuggling from the USA by a Pakistani agent was that of Ashad Pervez – a Pakistani-born Canadian – who was arrested in Philadelphia on 10 July 1987, charged with attempting to purchase and ship to Pakistan 25 tons of Maraging steel. The Pakistan Government has claimed it had nothing to do with the Pervez case.

In August 1985, according to *Der Stern*, a shipment of Maraging steel had been sent to Pakistan from West Germany. People with Pakistani connections arranged for 880 kilograms of the material to be sent to an address in Karachi. The steel went in the form of cylindrical rods whose diameter was exactly right to fit a type of centrifuge of a design known to Dr Khan.

Yet another nuclear smuggling case involving Pakistan came to light in the same week as the Pervez affair. Three individuals – two Americans and one from Hong Kong – were charged with smuggling sophisticated electronic equipment to Pakistan in 1982 and 1983, which could be used in Pakistan's nuclear-weapon programme.

There have been persistent reports that China has assisted Pakistan in its nuclear-weapon programme, both in operating the Kahuta uranium-enrichment plant and in nuclear-weapon design. It has also been rumoured that China has allowed Pakistan to test a nuclear explosive device at its nuclear test site in Lop Nor (see, for example, Spector, 1985 and 1987, and *Nucleonics Week*, 1986). China has consistently denied that it has given any nuclear assistance to Pakistan and it now seems that it is not aiding Pakistan's nuclear-weapon efforts. But President Reagan took the reports seriously enough to hold up approval of the 1984 US nuclear trade agreement with China for several months.

PAKISTAN'S NUCLEAR DELIVERY SYSTEM

As noted earlier, a modern nuclear weapon using uranium enriched to 93 per cent in uranium-235 should, with reasonably competent design, weigh no more than 500 kilograms. The Pakistan Air Force has a number of types of combat aircraft able to deliver such nuclear weapons; the US F-16 Fighting Falcon, the Chinese Q-5 Fantan, and the French Mirage 5PA3 and IIIEP. Some of Pakistan's Mirage IIIEPs carry the AM-39 Exocet air-to-surface missiles and could easily be modified to delivery nuclear weapons (the French Air Force itself operates Mirage IIIE aircraft carrying tactical nuclear bombs). It should be noted that commercial aircraft could be modified to operate as bombers carrying nuclear weapons. A Boeing 747, for example, equipped with suitable avionics would be as effective as a US B-52 strategic nuclear bomber.

HAS PAKISTAN TESTED A NUCLEAR EXPLOSIVE?

A number of reports in 1985 and 1986 claimed that Pakistan had tested the non-nuclear components of nuclear weapons. In particular, it was reported that in 1986 it had detonated two conventional assemblies of conventional high-explosive lenses (Carnegie Endowment for International Peace, 1988).

A useful test of an implosion device would consist of a dummy sphere of natural (or depleted) uranium surrounded with the shaped high-explosive charges needed to compress the uranium core. Such a test would, of course, not produce any fission reactions but would provide a great deal of useful information about the efficiency of the implosion technique.

One way of determining what is happening in the dummy core during such a non-nuclear test is to take flash X-ray pictures. According to an article in *Ny Technik* of 2 May 1986 by Christer Larsson and Jan Melin, Pakistan bought some flash X-ray machines, which can take flash photographs through solid masses, in 1982 from Scandiflash, a Swedish firm. It appears, however, that Pakistani scientists were unable to operate the machines.

The effectiveness of a test of the implosion technique using a dummy core can, however, be monitored using technologies other than flash X-rays. David Albright explains that

> electrical conducting pins placed at various depths within a hemisphere or three-quarters of a sphere of uranium could measure the arrival times of the high-explosive shock wave. By collecting this data, Pakistani scientists could compare arrival times of the shock wave with their calculated values and could also determine asymmetries in the shock wave (Albright, 1987b).

If the theoretical calculations turn out to be correct the scientists gain confidence in the nuclear-weapon design. If they do not, corrections can be made.

Pakistan could also test the design of its nuclear weapons without making a full-scale nuclear test by 'zero-yield' nuclear tests. In such a test, a small sphere of weapon-grade uranium is placed in the centre of the conventional chemical explosive lens system. The sphere is sufficiently small that, when the core is exploded, the nuclear fission yield produces an explosion of a power about the same as that of the conventional explosives in the lenses used to compress the uranium sphere. This amount of fission is, however, sufficient to be detected by radiation detectors placed around the assembly.

The detection of a burst of radiation, particularly neutrons, would show that effective implosion had taken place. Because the power of the explosion produced in such a low-yield test would be equivalent to that produced by the explosion of a few hundred kilograms of TNT, it could not be detected by seismic monitoring equipment outside Pakistan.

Pakistan is likely to avoid a full-scale nuclear test for political reasons. Such a step would, by removing all ambiguity from its nuclear-weapon programme, be extremely provocative. It would make it extremely difficult for any US Administration to resume economic and military aid to Pakistan; it would almost certainly trigger off a nuclear arms race with India; and it would subject Pakistan to a storm of international criticism. Given these negative consequences, and the fact that a full-scale test of an ordinary fission nuclear weapon is, in today's world, scientifically unncessary, it is doubtful whether the present

Pakistani Government would give its nuclear scientists permission to go ahead with such a test.

In summary, it is likely that Pakistan has tested the non-nuclear assembly for a nuclear weapon based on the implosion of weapon-grade uranium. It may also have exploded a nuclear device containing a small amount (a few kilograms) of weapon-grade uranium to check that the implosion produced some nuclear fission. In reply to a letter from President Reagan in 1984, President Zia ul-Haq gave assurances that Pakistan would not enrich uranium to more than 5 per cent in uranium-235. President Reagan accepted this assurance in spite of the evidence to the contrary. The American Administration is likely to want to continue to give aid to Pakistan. And this is why the US Government will do all it can to avoid having to admit that Pakistan has tested or assembled a nuclear weapon.

WHY AN ISLAMIC BOMB?

Pakistan's security policy is obviously greatly influenced by its relations with India. Pakistan will, for example, almost certainly refuse to ratify the Nuclear Non-Proliferation Treaty until India ratifies it. And the 1974 Indian nuclear explosion created a considerable incentive for Pakistan also to acquire the option to produce nuclear weapons. Bhutto, for example, called the Indian explosion a 'grave and serious threat'. A more grave one, he said, 'had not taken place in the history of Pakistan'.

Nuclear competition with India began before 1974. Pakistan's 'civil' nuclear programme in the 1960s and earlier, for example, was a reaction to India's civil programme. Senior Pakistani politicians and strategies continually demanded that the country should not fall too far behind India in nuclear technology. In 1966, for instance, Bhutto, then Pakistan's Foreign Minister, made his well-known remark that Pakistan would match India's nuclear programme even if the Pakistani people had 'to eat grass' to pay for it.

Since 1974, Pakistan's nuclear programme has acquired a significant momentum and it is common knowledge that the country has the capability to produce weapon-grade fissionable material. Because of the Indian nuclear explosion, there is much public support in Pakistan in favour of the country having nuclear-weapon force, or at least possessing the option rapidly to acquire nuclear weapons.

Nevertheless, Ashok Kapur, an expert on Pakistani affairs, points out that competition with India is not a sufficient explanation for Pakistan's nuclear programme.

> A careful study of Pakistani nuclear history and internal debates reveals the constant presence of multiple motivations, while the alleged threat from India has not always been of central importance in Pakistani strategic thinking. Generally speaking, after 1971 the primary motivation of the Pakistani leadership was to recover from the trauma of Pakistan's break-up and military defeat at Indian hands. In the second half of the 1970s Pakistani society had to adjust to the issue of military rule. By the late 1970s Pakistani security concerns had broadened to include the prospect of a permanent Soviet presence in Afghanistan (Kapur, 1985).

Some Pakistani strategists argue that Pakistani nuclear weapons will not only deter an Indian nuclear attack but will also cancel out India's superiority in conventional military forces. Others believe that Pakistan could acquire a nuclear strategic superiority, which would enable it to move on the Kashmir question, for example. Be this as it may, Pakistan's nuclear-weapon programme has undoubtedly greatly improved the country's prestige in the subcontinent. And this has had significant internal consequences, diverting attention from domestic economic and political problems.

So far as the consequences for the Middle East are concerned, an important factor is that some Arab countries have significantly assisted Pakistan in its nuclear programme, particularly financially. This may not give the financiers much control over the programme, but it ensures that Pakistan must justify its use of the money. This means that any Pakistani government would find it very difficult to reduce its nuclear efforts.

Pakistan's nuclear-weapon programme has greatly increased its prestige among Arab countries in the Middle East and the Gulf. Saudi Arabia, for example, is impressed by its 'development of the first Islamic and Arab nuclear bomb'. Pakistan has, in fact, become increasingly involved with some Arab countries, including involvement in Arab security issues.

The consequences of this increased prestige are described by Ashok Kapur:

> With a covert weapon component and a plausible economic justification, the country's activities have helped to mobilize both the security and development lobbies in Pakistani society, as well as to secure involvement of Pakistan's allies (the USA, Saudi Arabia, Libya and the People's Republic of China) in its economic development. By implying that its movement towards a nuclear explosive capability could be slowed down if its security and economic needs were satisfied, Pakistan has increased its leverage *vis-à-vis* its Arab, Chinese, European and North American suppliers. This may have facilitated the obtaining of economic aid from Libya, $3.2 billion from the USA and undetermined amounts of Saudi financing (Kapur, 1985).

Nevertheless, some Israeli strategists perceive a direct nuclear threat from Pakistan. This seems hardly justified given that the Middle East does not figure greatly in Pakistan's strategic thinking, despite its rhetoric and the obligations it has incurred because of its Islamic links. It is, of course, the consequences of Arab countries' assistance to Pakistan's nuclear-weapon programme that is of major concern to Israel – in particular, Libyan and Saudi Arabian involvement. Will this lead, some Israelis ask, to a future Pakistani government making nuclear weapons available to these countries?

Attention has been focused on the Saudi connection by its purchase of Chinese DF-3 Silkworm ground-to-ground missiles. With a 2,500-kilometre range, these missiles put

Saudi warheads within reach of Israel. Israeli worst-case analysts are bound to worry that in the future Saudi Arabia may put nuclear warheads on its Silkworms. The fact that in April 1988 Saudi Arabia announced its intention of ratifying the Nuclear Non-Proliferation Treaty is unlikely to allay Israeli fears, given Israel's contempt for the Treaty.

IRAN

Shah Mohammed Reza Pahlavi was a nuclear enthusiast, and under his regime, Iran had a very large nuclear programme. The Shah was anxious to establish Iran as the regional superpower – an ambition which led his government to purchase huge quantities of sophisticated conventional weapons. He probably also wanted the option to manufacture nuclear weapons. In fact, soon after India's nuclear test in May 1974, he said so openly, remarking that Iran should acquire nuclear weapons if any other country in the region did so.

The Shah must have known that the nuclear progress of Iraq, Iran's serious rival, and of Pakistan would give these countries the capability to produce nuclear weapon within ten years or so. To avoid the risk of being outdone, he would have wanted Iran to get into a position to acquire nuclear weapons rapidly. In any case, when a country with as much oil as Iran starts importing nuclear facilities, ostensibly for electricity generation, an interest in nuclear weapons is bound to be suspected.

Nevertheless, Iran ratified the Nuclear Non-Proliferation Treaty in 1970. It may well have done so in order to be able to buy peaceful nuclear technology, particularly nuclear-power reactors, abroad rather than to give up permanently any ambition to acquire nuclear weapons.

The construction of two large nuclear-power reactors, supplied by the West German firm Kraftwerk Union AG, began in the mid-1970s at a site 60 kilometres inland from the port of Bushehr. Each reactor was to have a generating capacity of 1,200 MWe. The site for another large nuclear-power reactor, to be supplied by the French firm Framatome, was prepared at Ahwaz. This was to be one of two reactors, each with a capacity of 900 MWe. According to the Shah's plan, these reactors were to be the beginning of a $30-billion nuclear-construction programme that would provide Iran with 20 or so nuclear-power reactors by the year 2000.

Arrangements were also made to guarantee supplies of low-enriched uranium fuel. To this end, Iran bought a 10 per cent share in EURODIF, a large uranium-enrichment plant built in France with capital from Belgium, France, Italy and Spain. Long-term uranium fuel contracts were made with France, West Germany, and the United States. Iran was also negotiating for a financial stake in another large enrichment plant, COREDIF, together with the same European partners (Boskma, 1974).

Bilateral nuclear co-operation agreements were entered into with Belgium, Canada, France, West Germany, Italy and the United States. And, in anticipation of his ambitious nuclear programme, the Shah sent thousands of people to these countries to be trained in a variety of nuclear specialities.

At the time of the Shah's deposition, there was an active nuclear group working at the Tehran Nuclear Research Centre. This Centre was well known to the international nuclear community, and a number of important nuclear conferences were held there. Many Iranian scientists had international reputations, and the Atomic Energy Organization of Iran (AEDI), founded in 1974, was a highly respected institution.

A small – 5 MWt – research reactor began operations at the centre in 1967. The reactor, fuelled with highly enriched uranium, was supplied by the American firm, General Atomic Corporation. Neither the reactor nor the fuel had any military significance. But, just before the Shah was overthrown, Iran took delivery of six lasers. These devices could be used to enrich uranium by the laser isotope separation technique and could therefore be of military significance. Because Iran had already arranged for copious supplies of low-enriched uranium for its nuclear-power programme, the purchase of the lasers could have been part of a plan eventually to produce highly enriched uranium for nuclear weapons.

AFTER THE REVOLUTION

When Ayatollah Ruhollah Khomeini came to power many nuclear scientists fled from Iran, but those that remained have not been idle (Segal, 1987). A second nuclear research centre – the Isafahan Nuclear Centre – has been started up, and according to *South* magazine, Iranian scientists have designed and built a small research reactor for the new centre.

The Khomeini regime stopped the construction of the partially built power reactors at Bushehr, disapproving of Iran's nuclear dependence on the Western powers. To break this dependence, the new regime signed bilateral agreements on nuclear collaboration with Pakistan and Argentina. Little is known about arrangements with Pakistan. A consortium – including the Argentinian firm Enace and Kraftwerk Union AG – has offered to move in and finish the construction of the Bushehr reactors. Apparently, there is no current plant to continue building the reactors at Ahwaz.

It is not clear whether the construction of the power reactors will start again while Iran is at war with Iraq. The site of the two reactors at Bushehr has been bombed by Iraq on at least seven occasions since the first attack on 24 March 1984. In a raid on 17 November 1987, a West German engineer was among the people killed. These events must discourage new nuclear construction which may well have to be postponed until after the war.

Argentina will also supply a new core for the research reactor at Tehran and the fuel for the reactor. The Argentinians have a comprehensive nuclear industry which has built its own research reactors, a reprocessing plant and a uranium-enrichment facility. Whether or not Argentina will eventually be prepared to help Iran build similar facilities to produce weapon-grade nuclear materials remains to be seen.

In the meantime, Iran intends to exploit its natural uranium reserves and has begun to develop a uranium mine in the Yazd province, where exploration indicates that there are some 5,000 tons of high-grade uranium ore.

Today's Iran claims the spiritual and political leadership of the world of Islam. It will not want any other Muslim state, like Pakistan, to obtain too much prestige by becoming Islam's only recognized nuclear-weapon power. The fact that the Ayatollahs are revitalizing Iran's nuclear programme should, therefore, give pause for thought.

The point will not be lost on Israel. American political scientist Robert Harkavy puts the point this way:

> Israel must take into account the possible formation of a 'grand coalition' of Islamic states, the hitherto chronic fragmentation of the Arab and Islamic 'world' notwithstanding. In past [Arab/Israeli] wars, nations such as Morocco, the Sudan, and Saudi Arabia have made token contributions of ground troops. In the future, however, Israel must at least take into account possible large-scale contributions from Iran and (if a shift either toward Islamic fundamentalism or Marxism should occur) perhaps Turkey (Harkavy, 1986).

Israeli strategists using, as most strategists do, worst-case analysis, will urge strong Israeli reactions to any signs of a renewed Iranian nuclear-weapon programme. And they will use the Iranian case to rationalize an Israeli nuclear arsenal containing a large number of nuclear weapons, including the most sophisticated types possible, and a long-range Israeli nuclear capability covering targets throughout the Islamic world.

7
Could the PLO Go Nuclear?

According to a British TV documentary, *The Plutonium Black Market*, transmitted by Channel Four on 30 October 1987, the Palestine Liberation Organization is in the market to buy fissionable material to construct nuclear explosives. The film showed evidence of a nuclear black market in Khartoum with the PLO as one of the potential customers. Another indication of PLO interest in acquiring nuclear material is the fact that it is mentioned in a document presented in the prosecution in Italy, started in 1984, of Glauco Partel, a member of an Italian arms smuggling ring. Partel is charged with having 'offered three atomic weapons for sale – weapons that Partel subsequently acknowledged did not exist – to a series of potential buyers with connections to Arab interests in the Middle East'. As Leonard Spector, who has thoroughly investigated the Italian case, says:

> This reference to the PLO is the only one in the dossier in connection with nuclear devices. While hardly conclusive evidence, it is at least an indication of PLO interest in the acquisition of nuclear arms, either directly or through the organization's ties with the Iraqi government (Spector, 1985).

The evidence that the PLO is interested in acquiring a nuclear device is admittedly scanty, but the potential for nuclear terrorism in a world containing rapidly increasing amounts of plutonium should not be underestimated. Peter Calvocoressi points out:

> Various types of belligerence fit uneasily into an international state system. With the growth of the power of the state it has become common to label as war only those kinds of organised violence which are conducted by a state . . . but there remain further kinds of belligerence, commonly disparaged as terrorism. This is a word to beware of. It has become a term of abuse used to excite prejudice and fuel unthinking reactions – which is all the more deplorable since terrorism does exist and has to be countered (Calvocoressi, 1987).

Calvocoressi's warning is particularly apt when dealing with the possibility that nuclear weapons will spread outside the control of sovereign states.

Terrorism is defined as criminal violence by minorities using 'coercive terror' for political ends. But one person's terrorist is another's freedom fighter. Gangs such as the Bader-Meinhof (often disaffected young middle-class people addicted to violence for its own sake) cannot be bracketed with organizations like the PLO which is seeking a national status for its people and has genuine reasons to despair of achieving its objectives by the normal political and diplomatic processes.

It must be emphasized that the PLO is a heterogeneous organization, consisting of a number of sub-groups. Some of these probably have much in common with other 'terrorist' groups in the Middle East, such as Islamic Jihad. But the PLO Fatah group resembles much more an army engaged in guerrilla warfare. Nevertheless, lacking legitimized national status, Fatah is neither protected by, nor subject to, international codes of behaviour and law. It therefore inevitably resorts to the methods of the terrorist.

Terrorism – i.e., sub-national violence for political reasons – can succeed in its purpose in politics, and violence can be made to pay. This lesson was, in fact, dramatically demonstrated in the Middle East when, after World War II, Jewish terrorists in Palestine, including the Irgun and Stern organizations, played a major role in driving out the British and establishing the State of Israel. Perhaps because of its successes, or perceived successes, terrorism is on the increase. As time goes on, sub-national groups are becoming more sophisticated and skilled. At the same time, wars, and society itself, are becoming more violent. We must expect that moral restraints on mass killing will weaken. That this is happening when access to nuclear material suitable for use in nuclear weapons is becoming easier is a cause of considerable concern.

As already explained, the design of a nuclear explosive, such as the bomb that destroyed Nagasaki in 1945, is no longer secret. Amory B. Lovins, for example, in an article in the scientific journal *Nature* in May 1979, summarized much of the necessary physics data showing that a competent nuclear physicist can find the relevant information in the open literature. As time goes on, more and more plutonium is being separated from spent reactor fuel elements. Plutonium will be increasingly transported worldwide on virtually all the main transportation systems – road, rail, sea, and air. It is most vulnerable to theft while it is being transported. Given enough plutonium and the details of design, it is not very difficult to construct a nuclear explosive device. Hence the fear that sub-national groups may acquire fissile material and build nuclear weapons.

Konrad Kellen, a senior staff member at the RAND Corporation, lists the principal nuclear activities that a sub-national group may indulge in as follows:

> the making or stealing of a nuclear weapon and its detonation; the making or stealing of a nuclear weapon for blackmail; the damaging of a nuclear plant for radioactive release; the attack on a nuclear-weapons site to spread alarm; the attack on a nuclear plant to spread alarm; the holding of a nuclear plant for blackmail; the holding off-site of nuclear plant personnel; the theft of fissionable material for blackmail or radioactive release; the theft or sabotage of things nuclear for demonstration purposes; and the attack on a transporter of nuclear weapons or materials (Kellen, 1987).

Any of these activities could, in theory, occur in the Middle East. But the most likely incidents are the manufacture of a nuclear explosive by a sub-national group, in order to detonate it or to use the threat to detonate for the purposes of blackmail. Currently, the groups most likely to want to acquire nuclear explosives are in the Middle East. If, for example, the Palestine Liberation Organization or, more accurately, a group within it, became convinced that Israel had a significant nuclear-weapon capability, it might feel compelled to try to acquire nuclear explosives for itself.

A PLO group is not the only Middle East candidate, however. A group of Muslim fundamentalists might also try to acquire a nuclear capability. In fact, just as sub-national groups tend to copy the behaviour of governments, particularly in military matters, so they tend to copy the behaviour of each other. If one group decides to manufacture or otherwise acquire nuclear explosives, others may well follow suit.

The proliferation of nuclear explosives to sub-national groups in the Middle East has considerable ramifications for world, as well regional, security. Not only could a nuclear sub-national group threaten to make nuclear attacks on countries outside the region, and be prepared to carry out the threat, but any use of nuclear explosives could escalate the conflict to a nuclear world war.

The dangers of such nuclear proliferation are, therefore, similar to those of the spread of nuclear weapons to countries. In practice, they are likely to be significantly greater. If a sub-national group decides to acquire the plutonium or enriched uranium that could be used to make a nuclear explosive, it would be under considerable pressure to construct and detonate the nuclear device before the authorities discover the fact. Given the extremely sensitive equipment now available to detect nuclear material, the group would have to assume that it would not be long before its nuclear device was discovered.

The explosion of a nuclear device is, however, not the only danger that would follow the illegal acquisition of fissionable material by a sub-national group. If some kilograms of the material were scattered – perhaps by a fire or conventional explosion – it would make a significant area of a city, for example, uninhabitable until it had been decontaminated, a process that could take a long time. Plutonium is a very toxic substance and particles – produced, for example, when it burns – inhaled into the lungs can cause lung cancer by the exposure of lung tissue to the radiation (particularly alpha particles) given off. The very possession by a sub-national group of significant amounts of nuclear material is, therefore, a threat in itself.

A government being blackmailed by a group known to have fissionable material would not need to be convinced that the group had the expertise to construct an effective nuclear explosive. Even if the device failed to produce a significant nuclear explosion it would almost certainly scatter nuclear material over a large area. And this would be threat enough.

Commenting on this possibility, American Ambassador Richard T. Kennedy told a US Congressional hearing on 24 March 1982 that although the objective of any group or individuals 'who, for political gain, sought to acquire a nuclear weapon' would hopefully be limited to extortion, 'nonetheless the extortion itself is something that cannot be tolerated; it is simply too dangerous a proposition' (US Department of Defense, 1987). It is this that makes nuclear threats special. The *mere possession* of fissionable material is likely to be enough to gain the ends of the group that acquires the material.

HOW EASILY COULD A SUB-NATIONAL GROUP MAKE A NUCLEAR EXPLOSIVE?

Politicians often dismiss the idea that sub-national groups could acquire nuclear explosives. They do so because they either do not understand the technicalities of the subject or they do not want to scare the public.

The nuclear industry also generally tries to play down the danger that fissionable material can be stolen. And it tends to argue that, even if a sub-national group somehow got hold of enough fissionable material, it would find it extremely difficult to design and construct a nuclear explosive. The industry, of course, has an axe to grind. The idea of illicit nuclear explosives is bad for business.

There is a good deal of disagreement about the ease with which a sub-national group could construct a nuclear weapon. Some misunderstandings arise because the nuclear explosives being considered use designs of very varied sophistication. The precise type of nuclear explosive under discussion is often not defined. Clearly, the less sophisticated the device, the easier it is to design and make.

A US Pentagon report on world commerce in nuclear materials, issued in November 1987, states:

> The technological expertise required to fabricate a crude [nuclear] explosive device has been the subject of controversy since the mid-1970s. Early concerns that a few individuals could construct a low-yield nuclear explosive device are still expressed, but they are recognized to have been exaggerated. To date, no such device is known to have been built or tested by terrorists. And the prevailing view among experts appears to be that fabrication of a bomb, even with high-grade weapons-usable material, would be extremely difficult but not impossible for a well-organized, well-financed terrorist group (US Department of Defense, 1987).

The report goes on to quote Thomas Schelling, of Harvard University:

> . . . it appears to require a group of significant size, high professional quality, and excellent organization and discipline to convert unauthorized or illicitly obtained nuclear materials into a usable weapon.

The Pentagon report is vague about the type of nuclear explosive it is referring to, but clearly what it defines as a 'usable weapon' is by no means the simplest. Furthermore, it makes clear that a sub-national group could fabricate a nuclear explosive device sophisticated enough to be described as a 'usable weapon'.

The Office of Technology Assessment (OTA) of the US

Congress also concludes that it would not be difficult for a substantial sub-national group to construct a nuclear explosive, if it got hold of fissionable material. In its publication *Nuclear Proliferation and Safeguards*, the OTA states:

> A small group of people, none of whom have ever had access to the classified literature, could possibly design and build a crude nuclear explosive device. They would not necessarily require a great deal of technological equipment or have to undertake any experiments. Only modest machine-shop facilities that could be contracted for without arousing suspicion would be required. The financial resources for the acquisition of necessary equipment on open markets need not exceed a fraction of a million dollars. The group would have to include, at a minimum, a person capable of researching and understanding the literature in several fields and a jack-of-all-trades technician. . . . There is a clear possibility that a clever and competent group could design and construct a device which would produce a significant nuclear yield (i.e. a yield *much* greater than the yield of an equal mass of high explosive) (OTA, 1977).

Similar conclusions are made by J. Carson Mark and his colleagues in their chapter in the book *Preventing Nuclear Terrorism* (1987). Because this group of authors includes experienced nuclear-weapon designers it is worth giving their conclusions in some detail. They say that, so far as crude nuclear devices (devices guaranteed to work without the need for extensive theoretical or experimental demonstration) are concerned:–

i) Such a device could be constructed by a group not previously engaged in designing or building nuclear weapons, providing a number of requirements are adequately met.

ii) Successful execution would require the efforts of a team having knowledge and skills additional to those usually associated with a group engaged in hijacking a transport or conducting a raid on a plant.

iii) To achieve rapid turnaround (that is, to make the

device ready within a day or so of obtaining the material), careful preparations extending over a considerable period would have to be carried out, and the materials acquired would have to be in the form prepared for.

iv) The amounts of fissile material necessary would tend to be large – certainly several times the minimum quantity required by expert and experienced nuclear-weapon designers.

v) The weight of the complete device would also be large – not as large as the first atomic weapons (about 4.5 tons), since these required aerodynamic cases to enable them to be handled as bombs – but probably more than a ton.

vi) The conceivable option of using oxide powder (whether of uranium or plutonium) directly, with no post-acquisition processing or fabrication, would seem to be the simplest and most rapid way to make a bomb. However, the amount of material required would be considerably greater than if metal were used.

vii) There are a number of obvious potential hazards in any such operation, among them those arising in the handling of a high explosive; the possibility of inadvertently inducing a critical configuration of the fissile material at some stage in the procedure; and the chemical toxicity or radiological hazards inherent in the materials used. Failure to foresee all the needs on these points could bring the operation to a close; however, all the problems posed can be dealt with successfully provided appropriate provisions have been made.

Groups within the PLO are, of course, well enough disciplined, sufficiently sophisticated, and have enough financial resources to carry out the operations needed to produce the sort of nuclear device which Mark et al. are considering. The truth of this statement is shown by the fact that the PLO effectively operates an arsenal which contains sophisticated conventional weapons.

According to Yezid Sayigh, the weapons used by the PLO against the Israelis in the 1982 Lebanon War included, in addition to a wide variety of light arms: RPG-7 anti-tank rockets; recoilless rifles (73-mm, 75-mm, 82-mm, 106-mm and 107-mm); anti-tank guns; Sagger anti-tank missiles; T-34 and possibly T-55 main battle tanks; BRDM-2 and BTR-152 armoured personnel carriers; anti-aircraft guns with calibres ranging from 12.7 mm to 100-mm; SA-7 surface-to-air missiles; SA-9 surface-to-air missiles; and self-propelled, radar-guided quadruple-barrelled 23-mm ZSU-23-4 guns (Sayigh, 1983). An organization that can effectively operate and maintain such an arsenal could construct a nuclear explosive of the type discussed by Mark et al.

It should be remembered that there are thousands of people today with knowledge of, and experience in, handling nuclear weapons. Some may be recruited by, for example, the PLO to assist in any planned nuclear operations. If no such people can be found who are sympathetic to the aims of the PLO, there will certainly be some who will co-operate for money.

Very crude nuclear explosives

The nuclear devices considered by Mark et al. are types similar to those dropped on Nagasaki and Hiroshima. But much cruder designs that will still give a powerful nuclear explosion are possible. These could produce nuclear explosions equivalent to the explosion of between 100 and 1,000 tons of TNT. They might yield several thousand tons, but are unlikely to yield 10,000 tons.

The crudest design could be easily constructed by a team of technicians (or a competent technician working alone) from, say, a sub-critical mass of plutonium. The plutonium need not be in metal form; plutonium oxide, for example, is more convenient and safer to handle. The plutonium oxide could be contained in a spherical vessel placed in the centre of a large mass of conventional high explosive. When detonated remotely by an electronic signal, the conventional explosive could compress the plutonium enough to produce some nuclear fission.

Such a device could be positioned so that, even if it did not produce any nuclear fission, the conventional explosion would widely disperse the plutonium. Under these circumstances, the plutonium would be scattered in the form of small particles, capable of being inhaled into the lungs. Because exposure to the alpha-particles given off when plutonium nuclei undergo radioactive decay can cause lung cancer, plutonium in particulate form is extremely toxic.

Even if the explosion from such a crude device was equivalent to the explosion of only a few tens of tons of TNT, it would completely destroy the centre of a relatively large city. For purposes of comparison, the largest conventional bomb used in World War II used about 10 tons of TNT; it was called the 'earthquake' bomb! An explosion equivalent to that of 100 tons of TNT exploded on the surface would produce a crater about 30 metres across.

In the words of Mason Willrich and Theodore Taylor:

> Under conceivable circumstances, a few people, possibly even one person working alone, who possessed about 10 kilograms of plutonium oxide and a substantial amount of chemical high explosive could, within several weeks, design and build a 'crude fission bomb'. By a 'crude fission bomb' we mean one that would have an excellent chance of exploding with the power of at least 100 tons of chemical high explosive. This could be done using materials and equipment that could be purchased at a hardware store and from commercial suppliers of scientific equipment for student laboratories (Willrich and Taylor, 1974).

It is often forgotten, or not taken into account, that a sub-national group would normally not need to predict the precise explosive power of any nuclear device it might make. Most likely purposes of the group would be satisfied if the device exploded with a significant nuclear yield. By contrast, nuclear-weapon designers employed by governments are required to produce devices which explode with specific yields, predictable to within relatively narrow limits. The weapons must also be reproducible. The military have very strict requirements for the weapons they are prepared to accept in their arsenals.

Another difference is that the weapons produced for a state's nuclear-weapon programme must satisfy the strictest possible safety conditions. The risk of accidents in which they are subjected to severe mechanical shocks and fires always exists. The weapons must be designed so that if the conventional high explosives are accidentally detonated, there is no nuclear fission. A sub-national group, on the other hand, would not have to take into account all conceivable accidents, although it would naturally want to make sure that it could handle its device safely.

The availability of plutonium

Although a sub-national group could choose to use either plutonium or highly-enriched uranium as the fissionable material for nuclear explosives, plutonium is increasingly the more likely option. Currently, roughly 17,000 kilograms of highly-enriched uranium is used throughout the world, mainly to fuel about 140 civilian research reactors in some 50 countries. But the development of new low-enriched uranium fuels for use in research reactors will sharply reduce the amount of highly-enriched uranium in circulation. The amount of plutonium available, however, will rapidly increase.

A sub-national group that, in the future, decides to manufacture a nuclear explosive is therefore most likely to try to steal or buy plutonium. As the amount of plutonium produced worldwide in civilian nuclear-power reactors and separated from spent reactor fuel elements in commercial reprocessing plants increase, it will become easier to obtain illegally.

The world's nuclear-power reactors have so far produced about 700,000 kilograms of plutonium. By the year 2000, this will have increased to 3 million kilograms. About 85 per cent of this will be produced in non-communist countries (Albright, 1987). In the year 2000, the world's nuclear power reactors will be producing about 160,000 kilograms of plutonium a year. By then, according to current plans, about 400 tons of plutonium will be reprocessed in the non-communist world. For purposes of comparison, the total amount of plutonium in the 50,000 or so nuclear weapons in today's nuclear-weapon arsenals is about 200 tons.

The spent fuel elements from nuclear-power reactors exported by the USSR to other communist countries are taken back to the USSR and stored there. If the elements are reprocessed the plutonium is kept in the USSR and not returned to the country that owns it. Plutonium produced in Soviet power reactors or Soviet-supplied reactors is, therefore, unlikely to be stolen.

Plutonium commercially reprocessed in non-communist countries is normally returned to its owners. As commercial reprocessing increases, a great deal of plutonium will be transported worldwide and will therefore be vulnerable to theft. Given the huge amount of plutonium in circulation, it will be virtually impossible to guard it effectively and prevent the theft of the relatively small amounts needed to produce nuclear explosives (only about 8 kilograms of the plutonium produced normally in commercial nuclear-power reactors are needed for a nuclear explosive).

THE NUCLEAR BLACK MARKET

With plenty of plutonium available and governments and sub-national groups willing to pay large sums of money for it, a flourishing nuclear black market is a very real possibility. Evidence collected by researchers for the TV documentary quoted earlier, *The Plutonium Black Market*, suggests that such a black market already exists. In the documentary, the former CIA Director, Admiral Stansfield Turner, said:

> I think there has been a black market in fissionable material for nuclear weapons and in the technology. In certain pieces of technology there obviously has been. The Pakistanis have tried to circumvent regulations on export of high-technology materials that are applicable to nuclear weapons and they have been successful in doing that.

When asked: 'Do you think there has been a black market in both plutonium and enriched uranium?', the

Admiral replied, 'Yes I do think there's been a black market.'

An anonymous arms dealer appearing in the documentary claimed to have been involved in the black market and said that its seeds were sown by Israel. Israel wanted to buy enriched uranium to produce nuclear weapons during the period before its nuclear facilities at Dimona were producing enough plutonium for nuclear weapons. In 1965, for example, Israel reportedly acquired some 240 kilograms of highly-enriched uranium which had been taken from a plant in the United States.

The arms dealer said that the black market was still operating and that its headquarters was in Khartoum, Sudan. He said that he had been offered 12 kilograms of plutonium, for example, which was sent to Khartoum and kept in a hangar at Khartoum airport, in drums lined with lead and concrete. The people bidding for the plutonium included representatives from Israel, South Africa, Pakistan, Iraq, Iran, Libya, and the PLO. Apparently, the plutonium was eventually bought by Iraq, for a price in excess of $5 million per kilogram.

The TV documentary claimed that in August 1987, a black-market consignment of enriched uranium was seized in Khartoum by the Sudanese police. Some of it was found in the house of a prominent businessman. The police arrested a ring of people involved. The Prime Minister, Sadiq Al Mahdi, appeared in the film and admitted the existence of a nuclear black market in the Sudan, saying that Khartoum was full of foreign agents wanting to buy uranium. He estimated that the uranium fetched about $3 million a kilogram. Other evidence about the nuclear black market in Khartoum was given by a retired member of Sudan's security service, Captain Assem Kabashi. He said that he had seen black-market uranium at Khartoum airport in 1986.

The Plutonium Black Market claims to have found evidence that, since 1980, six consignments of black-market nuclear material have been sold in the Sudan. Apart from the 12 kilograms of plutonium bought by Iraq, a further consignment of plutonium and one of weapons-grade uranium were sold in 1982. In 1987, three further consignments were sold in Khartoum – 6 to 10 kilograms of plutonium; a consignment of highly-enriched uranium; and a consignment of uranium, in which the concentration of uranium-235 was increased to 20 per cent.

With the increasing production of plutonium, the nuclear black market will flourish. When there is a surplus, the price will fall and sub-national groups will be more easily able to afford it.

Why have sub-national groups not already exploded a nuclear device? It seems almost certain that their leaders have at least thought about using nuclear and other weapons of mass destruction, such as those based on chemical and biological systems. Presumably, they have decided up to now that killing, or threatening to kill, large numbers of people indiscriminately, and contaminating a large area, would not further their ends. But this attitude may well change. The very presence of large amounts of plutonium and its frequent transportation may prove to be a temptation that cannot be resisted for long.

There is no doubt that credible evidence that the PLO had acquired a nuclear weapon, or a significant quantity of nuclear material suitable for use in a nuclear explosive, would produce serious social and political instability in Israel, perhaps amounting to mass hysteria. The possibility of provoking such extreme political disruption may, in fact, be the main objective of the PLO in acquiring nuclear explosives. This objective could, of course, be achieved without actually detonating a nuclear explosive device. Mere possession would be enough.

Ironically, it is the possible nuclear threat from the PLO that may be one significant reason for Israel's continuing nuclear ambiguity. Israeli political leaders may well fear that, if the PLO is faced with the disclosure of Israel's nuclear-weapon capability, it may feel compelled to obtain nuclear explosives itself. The PLO has, after all, acquired an arsenal containing many of the same types of conventional weapons as are in the possession of Israel.

PART III

THE ISSUE OF INTERNATIONAL CONTROL

8
The International Non-Proliferation Regime

Most governments believe that the further proliferation of nuclear weapons, particularly to unstable regions like the Middle East, is a serious threat to global and therefore to national security. Consequently, many countries are attempting to evolve national and international means of preventing or, at least, limiting such proliferation. This chapter will review such attempts. But, first, the consequences of a number of industrialized states moving into a plutonium economy will be discussed, because these will have considerable ramifications for efforts to control the spread of nuclear weapons.

THE PLUTONIUM ECONOMY

The situation in the Middle East, or for that matter in any other region, is complicated by the intimate relationship between military and peaceful atoms – in the words of Hannes Alven, the Nobel Prize-winning Swedish nuclear physicist, they are 'siamese twins'. The fact has to be faced that any country with a significant nuclear-power programme inevitably has the expertise and fissile material to produce nuclear weapons. All nuclear reactors produce plutonium as a by-product, and some of the skilled nuclear manpower required to operate and maintain nuclear-power reactors could be diverted to producing nuclear weapons. Moreover, as Pakistan, Israel, and Brazil have shown, Third World countries are well able to master the technology of uranium enrichment by the use of gas ultracentrifuges.

Nevertheless, the nuclear industries in the advanced countries, including the United States, France, West Germany, the United Kingdom, Italy, Japan and the USSR, are energetically trying to export nuclear facilities to Middle East countries. These industries need export orders to survive because of the current lack of domestic orders for nuclear-power reactors. Moreover, countries like France and Italy are exporting nuclear facilities and know-how to Arab countries in order to obtain guaranteed oil supplies.

The situation will become even more complicated if some of the advanced nuclear countries move into the plutonium economy, and current nuclear developments make it likely that they will. Some countries have made massive financial investments in nuclear power. Whereas nuclear-power reactors generate about 15 per cent of electricity in the United States and 10 per cent in the USSR, they generate 65 per cent of electricity in France, 60 per cent in Belgium and over 30 per cent in Sweden, Switzerland, Finland, Bulgaria and West Germany. And Japan, Czechoslovakia, the UK and East Germany are constructing enough nuclear-power reactors to generate a large fraction of their electricity by nuclear power in the near future.

By 2000, according to conservative estimates, the world's nuclear-power reactors will be generating about 600,000 megawatts of electricity (about twice the amount generated by nuclear reactors today). These reactors will also be producing about 160,000 kilograms of plutonium annually, and a number of countries are planning to reprocess the spent reactor fuel elements on a large scale to remove the plutonium from them. France, Japan, the UK, the USSR and perhaps the USA will have large commercial reprocessing plants by the early 1990s and Italy, West Germany and Belgium will have smaller but significant reprocessing plants.

As we have seen, reprocessing is done to produce plutonium for nuclear weapons, and also for fuel for breeder reactors. The preferred fuel for breeder reactors is plutonium rich in the isotope plutonium-239, plutonium ideal for the production of nuclear weapons.

Spent reactor fuel elements that are to be reprocessed will be transported from the countries operating nuclear-power reactors – and by 2000 there will be 35 or so of them – to the six or eight industrialized countries operating large commercial reprocessing plants. Separated plutonium will normally be transported back to the countries owning it. These operations will involve transporting large amounts of nuclear materials, including plutonium, by all forms of transport – road, rail, sea, inland waterways and air.

Spent reactor fuel elements are so radioactive as to be self-protecting. It would be extremely hazardous for people to handle them without large remote-handling equipment. But when the fuel elements have been reprocessed, and the plutonium separated from the radioactive fission products, the plutonium is in a form that can be relatively easily handled.

As emphasized in the previous chapter, if large-scale reprocessing takes place, there will, to say the least, be a considerable risk that sub-national groups will get hold of enough plutonium to make nuclear weapons. (Remember that only a few kilograms are needed to make a nuclear explosive and, by 2000, some 160,000 kilograms will be produced a year.) An obvious non-proliferation measure would to avoid reprocessing entirely and permanently dispose of spent reactor fuel elements. Current evidence suggests that the use of breeder reactors will not be economic for the production of electricity. Even if it could be shown that they are economic, the disadvantages of reprocessing, particularly the risk of the proliferation of nuclear weapons to nations and sub-national groups, far outweigh the advantages.

ISRAEL, PAKISTAN, AND IAEA SAFEGUARDS

One of the main tasks of the International Atomic Energy Agency, established in 1957, is to 'administer safeguards designed to ensure that special fissionable and other materials, services, equipment, and information made available by the Agency or at its request or under its supervision or control are not used in such a way as to further any military purpose' (Article III.5 of the IAEA Statute). The objective of safeguards is the timely detection of the diversion of significant quantities of nuclear material from peaceful nuclear activities to the manufacture of nuclear weapons or other nuclear explosive devices and the deterrence of such diversion by the risk of early detection. The IAEA safeguards system includes the application of measures for materials accountancy, supplemented by containment and surveillance.

IAEA safeguards begin to operate when an agreement is signed between the IAEA and the country owning the nuclear material under safeguards, which gives the Agency the right to make *ad hoc*, routine and special inspections. Inspectors are sent to the country to verify information that must be supplied to the Agency about the location, identity, quantity and composition of nuclear material subject to safeguards. Although IAEA safeguards are designed to *detect* the disappearance of nuclear material rather than to *prevent* such a disappearance, many exporters rely on the Agency to safeguard nuclear material produced in exported nuclear facilities.

Israel is an active member of the IAEA but has always been critical of the Agency for what it sees as its pro-Arab attitudes. Israel accepts IAEA safeguards on its small research reactor at Nahal Soreq, which was imported from the United States in the late 1950s. Initially, the safeguards on this reactor were set out in a bilateral US-Israel agreement; in 1966, responsibility for these safeguards was transferred to the IAEA.

Israel will not accept safeguards on its nuclear reactor and other facilities at the Dimona nuclear centre (including the fuel fabrication plant, the reprocessing plant, and the uranium-enrichment facility). Some US visits to the Dimona nuclear establishment were allowed after the US reconnaissance aircraft discovered it in 1963, but none have taken place since 1969. Even then, the US visitors were not allowed to see all the activities at Dimona and steps were taken to conceal the true purpose of the facilities. Other Israeli nuclear facilities not under safeguards include the uranium (UO_2) purification and uranium conversion (UF_6) plants and the heavy water plant at Rehovot.

Pakistan has several safeguards agreements with the IAEA relating to specific facilities, materials and equipment. These include a safeguards agreement covering the small PARR research reactor at Rawalpindi (in force since March 1962) and a trilateral (Pakistan/IAEA/Canada) agreement covering the KANUPP nuclear-power reactor at Karachi (in force since October 1969). A trilateral (Pakistan/IAEA/France) agreement covering the reprocessing plant to be supplied by France came into force in March 1976, but, as we saw in Chapter 6, France decided in 1978 to withdraw from the project.

Pakistani nuclear facilities not under international safeguards include the uranium enrichment plant at Kahuta, the uranium (UF_6) conversion plant at Dera Ghazi Khan, the fuel fabrication plant at Chashma, the reprocessing facilities at Rawalpindi, and the heavy-water plant at Multan. Generally, IAEA safeguards are in force when Pakistan has imported a nuclear facility and the exporter has demanded safeguards. When Pakistan can avoid safeguards it usually does so. Thus, facilities designed and constructed in Pakistan are not safeguarded.

THE 1968 NUCLEAR NON-PROLIFERATION TREATY (NPT)

The NPT is the main international instrument to prevent the spread of nuclear weapons. It has attempted to freeze the number of nuclear-weapon states at five – the United States, the USSR, the United Kingdom, France and China. Article I of the Treaty, which came into force in 1970, commits the nuclear-weapon parties (the USA, the USSR, and the UK) not to transfer nuclear weapons to, nor to assist in their manufacture by, the non-nuclear-weapon states. Article II pledges the non-nuclear-weapon states not to receive nuclear weapons or control over such weapons, and not to receive any assistance in the manufacture of nuclear weapons.

To verify compliance, the non-nuclear-weapon parties must, under Article III, sign agreements with the IAEA submitting all their nuclear activities to IAEA safeguards (safeguards on all nuclear facilities in a country are called full-scope safeguards). To encourage non-nuclear-weapon states to ratify the NPT, Article IV promises them co-operation and assistance in their peaceful nuclear programme.

Of the 160 or so countries in the world, 136 have ratified the NPT (in the Middle East, the important ratifiers include Egypt, Iran, Iraq, Libya, and Syria). At first sight, the NPT seems a strong treaty. But it is seriously weakened by the number of countries that either have nuclear weapons or could produce them in a short time and that have not joined the Treaty. These include France and China and also Israel, India, Pakistan, Brazil, Argentina, and South Africa.

The Treaty is also seriously weakened by the failure of the nuclear-weapon powers that have ratified it – the USA, the USSR and the UK – to fulfil their obligations under Article VI which legally binds them to take significant steps towards nuclear disarmament and halting and reversing the nuclear arms race. Instead of reducing their nuclear arsenals, these powers are continually increasing the number, and improving the quality, of their nuclear weapons and continuously developing technologies to support these weapons. By this behaviour they demonstrate that they believe that nuclear weapons have great political and military value. They cannot then be surprised when other countries follow their example and acquire a nuclear-weapon capability of their own. It is simply unconvincing to argue that nuclear weapons are good for some countries but not for others, and there are no grounds for stating that nuclear weapons deter war in Europe but will not deter war in other regions, like the Middle East.

The extension of the 1963 Test Ban Treaty into a permanent and comprehensive ban on nuclear-weapon tests is an example of a first measure that would show that the nuclear-weapon parties are taking their obligations under Article VI seriously and that would go some way towards strengthening the NPT. But to really strengthen the Treaty this must be rapidly followed by actual nuclear disarmament.

In 1995, a conference of the parties to the Treaty will be held to decide whether or not it should continue. Unless there is some change in the attitude of the nuclear-weapon powers the parties may well decide that the Treaty is not worth maintaining. This would be a serious blow to efforts to restrain the spread of nuclear weapons.

WHY ISRAEL REFUSES TO JOIN THE NPT

Israel demonstrated its attitude to the international non-proliferation regime when it bombed the Iraqi research reactor in June 1981, even though Iraq had ratified the NPT and the reactor and its fuel were under IAEA safeguards. Israel claimed that the reactor would have been used to produce weapon-grade fissionable material for nuclear weapons, and that neither IAEA safeguards nor other international non-proliferation measures were reliable enough to satisfy it that its national security was being protected.

Israel puts forward several reasons for not ratifying the NPT as outlined by Shalheveth Freier, an Israeli representative to the United Nations, in a statement to the UN First Committee on 2 November 1987:

> The NPT alone does not inhibit local wars, and local wars are the bane of the Middle East. For all the value of the NPT, let me tell you which of its deficiencies are pertinent in the Middle East context (Freier, 1987).

He went on to explain that the IAEA safeguards system used to verify compliance with the NPT was inadequate. To justify this he quoted a statement by the Director General of the IAEA, Hans Blix, on 11 December 1981:

> The safeguards do not, of course, reveal what future intention the State may have. It may change its mind on the question of nuclear weapons and wish to produce them despite possible adherence to the NPT. Neither such adherence nor full-scope safeguards are full guarantees that the State will not one day make nuclear weapons.

Statements such as this by no less a person than the head of the IAEA and certain statements by Arab politicians confirm the Israeli view that NPT safeguards are not sufficiently adequate to persuade it to join the NPT. Freier quoted Libyan leader Colonel Qadhafi, who was reported by Reuters as saying on 22 June 1987: 'The Arabs must possess the atom bomb to defend themselves, until their numbers reach one thousand million and until they learn to desalinate water and until they liberate Palestine'. As Freier pointed out, Libya is a party to the NPT. To these statements Freier added

> the qualifications attached by Syria and other Arab countries to their accession to the NPT. They qualify their commitment expressly by stating that their obligations under the NPT do not imply the recognition of Israel.

Freier also explained that the abrogation clause in the NPT was a major concern to Israel. Under Article X of the Treaty a party can at any time declare its withdrawal with three months' notice if it decides that 'extraordinary events' have occurred which 'it regards as having jeopardized its supreme interests'. The NPT, it is argued, allows a party to manufacture the components of a nuclear weapon, notify the IAEA and the UN Security Council that it is withdrawing from the Treaty, and then assemble its nuclear weapons.

Avi Beker, a political scientist at Bar-Ilan University in Ramat Gan, gives a good summary of Israel's attitude to the NPT:

> The fact that a majority of the world's states have accepted the NPT may create a false sense of security. The NPT was

helpful in enlisting those countries which had already accepted the political realities of international and regional order. The treaty cannot provide security in regions wherein certain countries are determined to change the political order by threatening the very existence of others. In such regions, a country's signature on the NPT cannot be regarded as conclusive proof of its nuclear innocence but, on the contrary, can be exploited as a strategy for the acquisition of nuclear arms. A system that is inadequate for controlling international transfers of nuclear equipment, materials, and technology generally is especially impotent in dealing with the Arab-Israeli conflict (Beker, 1986).

THE LONDON CLUB

The dangers of encouraging the spread of nuclear weapons by the uncontrolled export of nuclear technology have led to the major exporters trying to draw up common guidelines for nuclear exports. This effort is essentially an admission that the NPT is too fragile a treaty to prevent the spread of nuclear weapons to unstable regions such as the Middle East.

The seven major suppliers of nuclear materials and facilities – the USA, the USSR, the UK, France, West Germany, Canada and Japan – started meeting in London during 1975 to discuss ways of making the nuclear marketplace less chaotic. The original seven were joined by another eight suppliers – Belgium, Czechoslovakia, Italy, Switzerland, the Netherlands, Sweden, East Germany and Poland – and the group became known as the 'London Club'. The Club drew up a list of materials, equipment or technology, the so-called 'trigger list', which, when exported to any non-nuclear-weapon state, would 'trigger' IAEA safeguards. It also adopted guidelines for nuclear transfers requiring that countries receiving items on the trigger list should pledge themselves not to use them for the construction of nuclear explosives. Importers should also agree to provide effective physical protection for the materials supplied.

The guidelines set up by the London Club apply to facilities for reprocessing and uranium enrichment and also for the production of heavy water. They also apply to technologies directly transferred by the supplier or derived from the transferred facilities. The required safeguards apply also to any facility of the same type as that imported which is built indigenously during an agreed period, of some 20 years.

The guidelines are not a treaty; rather, they take the form of a gentleman's agreement. Each member gives an undertaking to the others to act according to the Club's guidelines when exporting nuclear materials, equipment or technology. A weakness of the Club is that it does not require importers of nuclear materials and facilities to adopt full-scope safeguards (i.e., safeguards on all its nuclear facilities rather than just on those imported), as the NPT does. And the Club's guidelines have not prevented some importers of nuclear technology and facilities from using them in nuclear-weapon programmes.

NATIONAL NON-PROLIFERATION MEASURES

Because of the weaknesses of the current international non-proliferation regime, a number of countries have unilaterally evolved national policies to try to prevent the further spread of nuclear weapons. These usually relate to the conditions under which nuclear materials and facilities are exported to countries, such as Israel and Pakistan, which have not ratified the NPT and which do not accept full-scope safeguards. Those Arab countries that have joined the NPT (Egypt, Iran, Iraq, Libya and Syria) are, of course, obliged to accept IAEA safeguards on all their nuclear facilities, whether imported or built indigenously.

Countries such as Australia, Sweden, Canada, and the United States will only export nuclear materials and facilities to countries other than the five nuclear-weapon countries recognized under the NPT if the importers accept full-scope safeguards or NPT membership. The United Kingdom also follows the guidelines of the London Club.

These strict nuclear exporters are at a commercial disadvantage when competing against others which are less strict, in particular France, Italy, and West Germany which require the application of safeguards to the exported items only.

The conditions under which the USA, for example, will co-operate with other countries in nuclear technology were laid down as long ago as 1954 in the US Atomic Energy Act. This specifies, in section 123, that non-nuclear-weapon states must permit IAEA inspection of all their nuclear facilities to ensure that no nuclear materials, technology, or equipment are being used in the manufacture of nuclear explosive devices. And section 129 states that non-nuclear-weapon states must not engage in activities involving plutonium or highly enriched uranium unless the US President determines that the country is taking steps to terminate such activities.

The US Nuclear Non-Proliferation Act of 1978 places statutory requirements on American nuclear exports. It requires full-scope safeguards. President Reagan, however, has said that the success of US non-proliferation policy depends on its ability to improve regional and local stability and to reduce the motivations of countries to acquire nuclear weapons.

> This calls for a strong and dependable United States, vibrant alliances and improved relations with others, and a dedication to those tasks that are vital for a stable world order (*Weekly Compilation of Presidential Documents* Vol. 17, No. 29, 20 July 1981).

Although the President confirmed that the policy of his Administration would include the inhibition 'of sensitive transfers' of nuclear technology, equipment and materials, he said that becoming 'a reliable nuclear supplier' was essential to US non-proliferation goals. As Warren Donnelly points out:

> The Ford and Carter Administrations favoured uniform action, treating all countries the same, whereas, in the

name of 'new realism' the Reagan Administration would tailor its decisions to individual circumstances (Donnelly, 1986).

The US Government has taken a number of other measures to prevent the spread of nuclear weapons. These typically take the form of restrictions on foreign aid. The 1976 Symington Amendment, for example, prohibits US aid to any non-nuclear-weapon state that imports or exports equipment or technology for the enrichment of uranium unless that country accepts full-scope safeguards. The US President may waive the ban if he certifies that he has received reliable assurances that the country concerned will not acquire nuclear weapons or assist others to do so and that the ban on aid would have a serious adverse effect on 'vital US interests'.

The Symington Amendment, put forward by former Senator Stuart Symington, is attached as an amendment to the Foreign Assistance Act. Although provoked by the 1975 West German agreement to sell Brazil a whole range of nuclear facilities, including a uranium-enrichment plant, it was primarily aimed at dissuading Pakistan from developing nuclear weapons.

The 1977 Glenn Amendment, introduced by Senator John Glenn, bans the provision of US aid to countries that import plutonium reprocessing technology or equipment. A subsection was added to the Amendment in 1981 prohibiting US aid to any non-nuclear-weapon state that receives, detonates, or transfers a nuclear explosive device. The President can waive the amendment by certifying that cutting off aid to the country concerned would seriously jeopardize America's non-proliferation policy or national security interests. But he cannot waive the subsection for more than 30 days unless Congress enacts a new law authorizing the waiver.

The 1985 Solarz Amendment, introduced by Congressman Steven Solarz, bans US aid to any country found by the President to have imported or attempted to import material, equipment, or technology illegally exported from the USA that could significantly assist it in the manufacture of a nuclear explosive device, if the President finds that the export was intended to be used for this purpose. The President may waive this ban if he determines that banning US aid would seriously harm US non-proliferation efforts or jeopardize the common defence and security.

US law therefore requires a ban on economic and military aid to Pakistan if Pakistan should announce that it has nuclear weapons or if the US should determine that it has them. This prohibition does not, however, legally apply to Israel because the US Government accepts the official Israeli position that Israel is not manufacturing nuclear weapons. As Warren Donnelly points out:

> To continue US aid if Israel declared [that it possessed] nuclear weapons would be seen by some countries as evidence that US support for Israel outweighs US non-proliferation policy, and would weaken US efforts to keep Pakistan and India away from such weapons (Donnelly, 1986).

Clearly, there is a difference between US nuclear policy to Pakistan and that to Israel. The pressure, from Congress and the White House, on Pakistan to prevent it from manufacturing, and even more from testing, nuclear weapons is considerable and high-profile, whereas no such visible pressure is put on Israel to stop its nuclear-weapon programme. This difference clearly reflects the special relationship between Israel and the USA. But it also shows very clearly that the US is prepared to modify its non-proliferation policy for the sake of other foreign considerations.

INTERNATIONAL CONTROLS: HAVE THEY WORKED IN THE MIDDLE EAST?

International controls have not prevented Israel and Pakistan, neither of them members of the NPT, from acquiring a nuclear-weapon option. In the process, both countries have flouted legal obligations. The weakness of international nuclear controls is demonstrated by the ease with which Israel was able to get around restrictions on its use of heavy water (see Chapter 4).

Illicit nuclear assistance to countries in the Middle East region again hit the headlines in early 1988 because of the activities of a West Germany company, Transnuklear, Transnuklear has been accused of selling plutonium to Libya and Pakistan, possibly through an unnamed intermediary country. It also stands accused of bribing Belgian officials at the Mol nuclear establishment near Antwerp, who are said to have been given DM 21 million to re-label 2,400 drums containing nuclear material.

If Transnuklear did ship plutonium to Pakistan it will not be the first time that Pakistan has been involved in shady nuclear deals. Pakistan has, if anything, been more successful at nuclear smuggling than Israel or any other country. As has been described in Chapter 6, much of the information and technology that Pakistan has used in its nuclear-weapon programme has been clandestinely acquired abroad, from countries in Scandinavia, Europe, China, and North America.

Pakistani and Israeli experiences demonstrate that determined people can get hold of the materials and technology to make nuclear weapons. And when they do, the international community rarely takes strong action. Governments and agencies sometimes set up inquiries but they seldom get to the bottom of the incidents. For example, EURATOM set up an inquiry into the 200 tons of uranium from Union Minière du Haut Katanga that ended up in Israel in 1968. The EURATOM official who authorized the sale of this uranium has admitted that the inquiry team 'agreed to cover up the loss. It was an embarrassment and no government had an interest to publicize it.'

9 Reducing the Importance of Nuclear Weapons in the Middle East

WHAT CAN OTHER COUNTRIES DO?

The prevention of the further proliferation of nuclear weapons in the Middle East should be a main aim of all governments. For a start, they should face the facts and stop supporting the myths perpetuated by the Israeli and Pakistani Governments that they do not have nuclear-weapon programmes. The practice subscribed to by virtually all governments is one of officially defining a nuclear-weapon state as one that has tested a nuclear explosive device. A country that has manufactured the components for nuclear weapons is not officially recognized as a nuclear-weapon state until it has tested a nuclear device. And, as has already been explained, nuclear scientists in today's world would be confident that ordinary nuclear weapons would work effectively and as predicted without testing.

The reasons why governments avoid acknowledging obvious nuclear facts is that it would be embarrassing for them to do so. If the Arab countries, for example, were openly to admit that Israel has a large and increasing nuclear-weapon force they would be under considerable domestic pressure to react. If the United States were to admit it Congress might cut off economic and/or military aid.

Until the true nuclear status of Israel and Pakistan is officially recognized there can be no useful debate about the consequences of nuclear-armed Middle Eastern countries for regional or world security. And, until the security risks are openly recognized, governments will continue to avoid taking action to reduce the danger of proliferation in the region.

Stopping the spread of nuclear weapons in the Middle East will not be an easy task. If Israel continues to increase the number and improve the quality of its nuclear weapons, the major Arab countries, and perhaps the PLO, will be under increasing pressure to acquire nuclear weapons. It is therefore essential that the Arab states that have joined the Non-Proliferation Treaty – particularly Egypt, Iraq, Libya, Syria, Jordan, and Saudi Arabia – should be encouraged to stay in the Treaty and fulfil all their obligations under it. This requires that the Treaty be strengthened. In particular, the nuclear-weapon parties should begin to fulfil all their obligations under the Treaty. But, given its general attitude to international arms control treaties and its specific objections to the NPT, Israel will not easily be persuaded to sign the Treaty.

A NUCLEAR WEAPON-FREE ZONE IN THE MIDDLE EAST

Each year since 1974 the United Nations General Assembly has adopted a resolution entitled 'Nuclear Weapon-Free Zone in the Region of the Middle East'. Originally, the resolution was proposed by Egypt and Iran. Israel, which had not been consulted beforehand about the resolution, as is the normal diplomatic practice, did not support it and abstained during the vote.

In a speech to the General Assembly on 30 September 1975, however, Israel's Foreign Minister Allon proposed that all the important states in the Middle East should hold consultations about a nuclear weapon-free zone in the region. In his words:

> Israel supports the proposal for a Nuclear Free Zone in the Middle East and will be ready to enter into negotiations with all states concerned in order to attain this objective. By negotiations we mean a process or intergovernmental consultations similar to that which preceded the adoption of

> the Treaty of Tlatelolco and other international instruments of like character. We do not think that so grave a matter can be settled by correspondence through the Secretary General.

The Allon speech represented a change of Israeli policy. Previously, Israel had insisted that any arms control measures in the Middle East should include conventional weapons as well as weapons of mass destruction. Allon surprisingly dropped this linkage.

At that time Iran, a sponsor of the 1974 UN resolution, was an ally of Israel. And as Efraim Inbar points out:

> By fall 1975 Allon succeeded in convincing Rabin to support the Iranian step and to allow him to propose a nuclear weapon-free zone. The argument that it was desirable to please Iran, Israel's main source of oil at that time and a regional ally, appealed to Rabin. At that time, Israel was somewhat concerned about a possible change in Iran's policy toward it (Inbar, 1986).

Allon and Rabin would have known, of course, that it was extremely unlikely that a nuclear weapon-free zone in the Middle East would be negotiated. But they would also have known that the Allon speech would considerably reduce US suspicions about Israel's nuclear-weapon plans.

Since the 1979 Iranian revolution, Egypt has sponsored the UN resolution on a nuclear weapon-free zone in the Middle East alone. And since 1980 Israel no longer abstains on the vote so that there is a General Assembly consensus and the resolution is adopted without a vote.

Typically, the UN resolution reads:

> Urges all parties directly concerned to consider taking steps required for the implementation of the proposal to establish a nuclear weapon-free zone in the region of the Middle East and, as a means of promoting this objective, invites them to adhere to the NPT; calls upon all countries of the region that have not done so, pending the establishment of the zone, to agree to place all their nuclear activities under

IAEA safeguards; to declare their support for establishing such a zone, and depositing those declarations with the Security Council; and not to develop, produce, test or otherwise acquire nuclear weapons or permit the stationing on their territories, or territories under their control, of nuclear weapons or nuclear explosive devices.

What is normally meant by the term 'nuclear weapon-free zone'? In plain language, a nuclear weapon-free zone, in its pure form, is an area from which nuclear weapons are totally excluded. The area covered may be part of a country, a whole country or a number of countries in a region. Nuclear weapons would never be allowed in such a nuclear weapon-free zone under any circumstances. For example, a nuclear-weapon power would not be allowed to land military aircraft carrying nuclear weapons in any country which was part of a nuclear weapon-free zone.

A legal definition of a nuclear weapon-free zone is given in a UN Resolution of 11 December 1975. It is a zone

> recognized as such by the UN General Assembly, which any group of states, in the free exercise of their sovereignty, has established by virtue of a treaty or convention whereby: (a) the statute of total absence of nuclear weapons to which the zone shall be subject, including the procedure for the delimitation of the zone, is defined; and (b) an international system of verification and control is established to guarantee compliance with the obligations derived from that statute.

For the purposes of a nuclear weapon-free zone, a nuclear weapon can be defined as follows (as it is in the 1968 Treaty of Tlatelolco). A nuclear weapon is any device which is capable of releasing nuclear energy in an uncontrolled way and which has characteristics that are appropriate for use for warlike purposes. An instrument that may be used for the transport or propulsion of the device is not included in the definition if it is separable from the device and not an indivisible part of it.

As noted earlier in this study, a nuclear weapon is, for

arms control purposes, legally not a nuclear weapon until the component containing the plutonium or enriched uranium is put into it. If this component is stored outside the weapon, it is not legally a nuclear weapon. This means that a country describing itself as not having nuclear weapons could have made all the components. It only becomes a nuclear-weapon power when it assembles the weapons, which could, of course, be done very quickly. This difficulty of legal definition, which seems inevitable, is a weakness of all treaties banning nuclear weapons from an environment.

It should be remembered that the NPT commits the non-nuclear-weapon parties not to receive nuclear weapons, not to manufacture or otherwise acquire them, and not to seek or receive any assistance in their manufacture. If a region contains a group of countries that have ratified the NPT it is not necessarily a nuclear weapon-free zone. The NPT allows the deployment of nuclear weapons in a party to the Treaty provided that the weapons remain under the control of the nuclear-weapon state that owns them. A nuclear weapon-free zone, on the other hand, does prohibit such deployment and is, therefore, a more comprehensive ban on nuclear weapons than the NPT.

Nuclear weapon-free zones have been established in a number of uninhabited regions – the Antarctic (by the 1959 Antarctic Treaty), outer space (by the 1967 Outer Space Treaty), and the sea-bed and the ocean floor (the 1971 Seabed Treaty). These nuclear weapon-free zones were relatively easy to negotiate because there is virtually no military interest in deploying nuclear weapons in the regions concerned, and they are all uninhabited.

The only inhabited regions to be declared nuclear weapon-free zones are Latin America and the South Pacific. The 1968 Treaty of Tlatelolco prohibits the testing, use, manufacture, production or acquisition by any means, as well as the receipt, storage, installation, deployment and any form of possession, of any nuclear weapons by Latin American countries. The 1986 Treaty of Rarotonga prohibits the manufacture or acquisition by other means of any nuclear explosive device, as well as possession or control over such device, by the parties anywhere inside or outside the South Pacific zone.

In a nuclear weapon-free zone in the Middle East the countries concerned would, in general terms, presumably sign and ratify a treaty establishing the zone. The parties would commit themselves to use any nuclear facilities and material under their jurisdiction exclusively for peaceful purposes. They would agree not to produce, or to acquire in any way, or to test nuclear weapons, and to prohibit the receipt, installation, storage, or deployment of any other country's nuclear weapons.

Israel has said that it is willing to negotiate a nuclear weapon-free zone in the Middle East. Negotiations for such a zone, and the local mutual arrangements that would have to be included in it, Israeli officials argue, would help prevent the outbreak of local wars in the region. In the words of Shalhaveth Freier to the UN First Committee: 'It is inconceivable that states negotiate a nuclear weapon-free zone and mutual arrangements, and continue to contemplate intermittent, occasional war'. The commitment to a nuclear weapon-free zone would, according to Freier, preclude the settlement of disputes by even conventional warfare. It will be remembered that Israel's view is that the NPT does nothing to inhibit local wars.

The crucial Israeli demand is that a nuclear weapon-free zone in the Middle East should be *negotiated*. The Egyptian proposal for such a zone, spelt out in the UN Resolution, does not require any negotiation between the Arab countries and Israel. What is required is simply that the countries concerned join the NPT, in the meantime place all their nuclear facilities under IAEA safeguards, and prohibit other countries stationing nuclear weapons on the territories of the countries in the zone. In other words, the resolution requires those countries in the Middle East that have not ratified the NPT to do so unilaterally, adding only the prohibition of third countries deploying nuclear weapons in the region. Israel is, of course, one country that would have to ratify the NPT, but a number of Arab countries would also have to do so.

The membership of the League of Arab States (excluding

the PLO) consists of Algeria, Bahrain, Djibouti, Iraq, Jordan, Kuwait, Lebanon, Libya, Mauritania, Morocco, Oman, Qatar, Saudi Arabia, Somalia, Sudan, Syria, Tunisia, the United Arab Emirates, the Yemen Arab Republic, and the People's Democratic Republic of Yemen. Egypt's membership was suspended in 1979 at the time of the peace treaty with Israel. Of these countries, the following have not ratified the NPT: Algeria, Bahrain, Djibouti, Kuwait (has signed but not ratified), Mauritania, Oman, Qatar, Saudi Arabia, and the United Arab Emirates. The following countries still have to complete the Safeguards Agreement with the IAEA required by the NPT: Somalia, Syria, Tunisia, the Yemen Arab Republic, and the People's Democratic Republic of Yemen.

The ratification process does not require any negotiation. The other requirements of the UN resolution could be fulfilled by declaration. When Israel voted in 1980 and thereafter in favour of the Egyptian draft resolution it reserved its position on the modalities, stating that the principle of regional negotiation must apply.

In 1980, Ambassador Arieh Eilan, Israel's representative to the UN First Committee, introduced a resolution calling on the countries in the Middle East 'and non-nuclear-weapon states adjacent to the region' to convene a conference to negotiate, 'regardless of their political differences and without prejudice to any political and legal claim', a multilateral treaty setting up a nuclear weapon-free zone in the region. The Arab states unanimously rejected the Israeli resolution. The Arabs were not prepared to accept 'negotiated regional arrangements'. Commenting on this, Israeli scholar Avi Beker writes:

> This is an unfortunate reality of the Middle East. Arab states refuse to participate in a multinational, treaty-writing conference which would require the acceptance of Israel as a legitimate Mideast entity and might imply the beginning of formal interstate diplomacy. The Iraq statement went even further by referring to the 'Zionist entity' (a usual Iraqi practice) and in fact denying Israel's right to even be at the committee (Beker, 1986).

Given the power of the Arab vote at the UN, Israel withdrew its resolution.

When they ratified the NPT, some Arab countries added the qualification that their obligations under the Treaty did not imply recognition of the State of Israel. The Arab countries are therefore unwilling to imply recognition of Israel by sitting down with it to negotiate a treaty. And, of course, it is, at least partly, in order to achieve recognition that Israel wants such a conference.

The Israelis believe, however, that there are other reasons why the Arab countries refuse to negotiate a nuclear weapon-free zone. In the words of Freier to the UN First Committee on 2 November 1987:

> The Arab refusal . . . must needs be interpreted as a desire to maintain the option of waging war against Israel also in the future. The NPT, as you know from all current wars, presents no impediment to such a situation. Also, the Arab refusal to entertain mutual arrangement with Israel within a nuclear weapon-free zone must necessarily be interpreted as an intention to avail themselves of the licences open to them under the NPT (Freier, 1987).

Freier was referring to the fragility of the NPT safeguards system. The Israelis believe that a non-proliferation arrangement for the Middle East must contain a control and verification system that is much stronger than that provided by the IAEA. IAEA safeguards may work in regions that are stable and in which there are no immediate conflicts, but more stringent safeguards are required in unstable and conflict-ridden regions like the Middle East. Moreover, the acts of negotiating such safeguards and accepting the mutual arrangements that they would require may well have a stabilizing influence.

Those who argue for strong verification provisions often point to the Treaty of Tlatelolco as a model. Thus, the Israeli Foreign Minister Itzhak Shamir, speaking in the UN General Assembly on 1 October 1980, after saying that the NPT 'cannot effectively prevent' nuclear-weapon proliferation in the Middle East, went on:

> The only genuine way to remove the nuclear threat in the Middle East can be found in the establishment of a nuclear-weapon-free zone, freely and directly negotiated among the countries of the region and based on mutual assurances, on the pattern of the Tlatelolco Treaty for Latin America.

The Tlatelolco Treaty set up a special permanent body to ensure compliance called the Agency for the Prohibition of Nuclear Weapons in Latin America, with headquarters in Mexico City. The Council of this Agency has the power to carry out special inspections.

When so requested 'by any Party which suspects that some activity prohibited by this Treaty has been carried out or is about to be carried out, either in the territory of any other Party or in any other place on such latter Party's behalf, the Council shall arrange for the special inspection'. The parties undertake 'to grant the inspectors carrying out such special inspections full and free access to all places and all information which may be necessary for the performance of their duties and which are directly and intimately connected with the suspicion of violation of this Treaty'. In addition to this special verification system, each party to the Treaty must negotiate an agreement with the IAEA for the application of IAEA safeguards to its nuclear activities.

Undoubtedly, the provision for inspection by challenge by a special local body, in addition to IAEA safeguards, makes for a verification system that is considerably stronger than that provided by the NPT. Although it is no more able to prevent the diversion of fissionable material from peaceful to military purposes than the ordinary IAEA safeguards system, it makes it easier to investigate in a timely way any diversion which has taken place. This is why some Israelis argue that a Middle Eastern nuclear weapon-free zone must contain a verification system similar to that provided by the Treaty of Tlatelolco – one that provides a reliable system of inspection by challenge.

Another feature of the Treaty of Tlatelolco which many Israelis find attractive is Additional Protocol II. This obligates the nuclear-weapon states to respect the statute of demilitarization of Latin America, not to contribute to acts involving violations of the Treaty, and not to use or threaten to use nuclear weapons against the parties to the Treaty. There is no such provision in the NPT.

A nuclear weapon-free zone in the Middle East would be an effective way of strengthening the non-proliferation regime. It is, however, difficult to be as optimistic as some Israelis are that a Middle East nuclear weapon-free zone modelled on the Treaty of Tlatelolco would significantly reduce the likelihood of the outbreak of local wars in the region.

The support of Israel for a nuclear weapon-free zone, coming after the 1979 Egyptian-Israeli peace treaty, is a gesture of good-will towards Egypt. But would Israel give up its nuclear weapons to join a nuclear weapon-free zone? Many doubt that it would, at least until there has been a significant move to a comprehensive peace between Israel and the Arabs.

In any case, the prospects for the establishment of a nuclear weapon-free zone in the Middle East are, in the short term, bleak. In the words of Avi Beker:

> The paradox is that the conflictual nature of the Middle East makes the NPT and the IAEA safeguards unreliable while at the same time it compels the adoption of unique methods of cooperation among the hostile parties. However, a dramatic development can always occur and make the nuclear weapon-free zone idea more feasible. The governments of the Middle East will have to realize that the irradiated particles that comprise nuclear fallout cannot distinguish between Jew and Arab, between Moslem and Christian (Beker, 1986).

An embargo on nuclear exports to the Middle East?

In the meantime, countries should look to the conditions under which nuclear exports to the Middle East take place. The situation is complicated by the fact that some European countries are hostages to Arab oil, and are therefore tempted to agree to supply nuclear materials and facilities with inadequate international safeguards in order to obtain guaranteed long-term supplies of oil. France and Italy, for example, have been prepared to export nuclear facilities to Iraq and secure oil supplies.

Some of those who argue that there can be no adequate international safeguards in such unstable regions as the Middle East, demand a total embargo on all nuclear exports to the region. This would, of course, be the ideal policy.

It is most important that all Middle Eastern countries should realize that there are no rational military uses for nuclear weapons. Such a recognition would reduce the pressure to acquire them. The five established nuclear-weapon powers – the United States, the USSR, the United Kingdom, France and China – could help this process by behaving as if they believed this fact of the nuclear age. Under no circumstances should they deploy their nuclear weapons in the region.

The nuclear-weapon powers can best demonstrate that nuclear weapons have no military utility by negotiating far-reaching nuclear disarmament. This would create an international climate in which nuclear-weapon proliferation is least likely to take place.

IMPORTANCE OF A PEACE SETTLEMENT

If the conditions of peace under which Israel would give up its nuclear weapons were ever achieved, the need to establish a nuclear weapon-free zone in the Middle East would, some say, disappear. It is undoubtedly true that a peace settlement is as important a move as better control over nuclear weapons and nuclear-weapon proliferation – if not more so. But until it is achieved an effort should be made to denuclearize the Middle East, particularly because this would be likely to assist the peace process.

Aharon Cohen has pointed out the core of the problem:

> Any solution to the problem of Israel-Arab relations must be tested by its ability to provide at one and the same time an

> acceptable answer to two basic and interrelated problems:
> 1. Recognition by the Arabs of the state of Israel, and the integration of Israel into the political totality of the area, which is basically Arab.
> 2. Satisfaction of the personal and the national-political rights of the Palestinian Arabs – permanent population and refugees alike (Cohen, 1970).

These words were written in 1970 but are equally true today. If the Arab states and Israel sit down to negotiate a solution to the problem of the proliferation of nuclear weapons in the Middle East this act could itself satisfy the first of Cohen's conditions. Be this as it may, few will deny that the need to prevent such proliferation is one of the Middle East's most vital and urgent tasks.

Appendix I
Types of Nuclear Weapons

To understand the role of plutonium and enriched uranium in nuclear weapons some knowledge of their design is desirable. A brief description of nuclear weapons follows.

A-BOMBS

The basic nuclear weapon is the fission bomb, or A-bomb (A for atomic) as it was first called. A fission chain reaction is used to produce a very large amount of energy in a very short time – roughly a millionth of a second – and therefore a very powerful explosion.

The fission occurs in an isotope of a heavy element – either uranium or plutonium. Specifically, the atomic bombs built so far have the isotopes uranium-235 or plutonium-239 as the fissile material. A fission occurs when a neutron (one of nature's elementary particles) enters the nucleus of an atom of one of these materials, which then breaks up or 'fissions'. When a fission occurs a large amount of energy is released; the original nucleus is split into two radioactive nuclei, the fission products; and two or three neutrons are released. These neutrons can be used to produce a self-sustaining chain reaction. A chain reaction will take place if at least one of the neutrons released in each fission event goes on to produce the fission of another heavy nucleus.

A critical mass exists for uranium-235 and plutonium-239 – the smallest amount of the material in which a self-sustaining chain reaction, and hence a nuclear explosion, will take place. The critical mass depends on the nuclear properties of the material used for the fission, whether it is uranium-235 or plutonium-239; the density of the material (the higher the density, the shorter the average distance travelled by a neutron before causing another fission and therefore the smaller the critical mass); the purity of the material (if materials other than the one used for fission is present, some neutrons may be captured by their nuclei instead of causing fission); and the physical surrounding of the material used for the fission (if the material is surrounded by a medium like natural uranium, which reflects neutrons back into the material, some of the neutrons may be used for fission which would otherwise have been lost, thus reducing the critical mass).

The critical mass of, for example, a sphere of pure plutonium-239 metal, in its densest form, is about 10 kilograms. The radius of the sphere is about 5 centimetres, about the size of a small grapefruit. If the plutonium sphere were surrounded by a natural uranium neutron reflector, about 10 centimetres thick, the critical mass would be reduced to about 4.4 kilograms, a sphere of radius of about 3.6 centimetres, about the size of an orange. A 32-centimetre thick beryllium reflector reduces the critical mass to about 2.5 kilograms and a sphere of radius of 3.1 centimetres.

Using a technique called implosion, in which conventional chemical explosives are used to produce a shock wave which uniformly compresses the plutonium sphere, the volume of the sphere can be slightly reduced. Its mass is then increased. If the original mass of the plutonium is slightly less than critical it will, after compression, become slightly greater than critical and a nuclear explosion will take place.

Using implosion, a nuclear explosion could, with a good modern design including a good reflector, be achieved with some 2 or 3 kilograms of plutonium. A 2-kilogram sphere of plutonium metal would have a radius of about 2.8 centimetres, smaller than a tennis ball. The trick is to obtain very uniform compression of the sphere.

In such a design, the plutonium would be surrounded by

a spherical shell, called a tamper, made from a heavy metal like natural uranium. The tamper has two functions. First to reflect back into the plutonium some of the neutrons which escaped through the surface of the plutonium core in order to minimize the mass of plutonium needed. Second, because the tamper is made of heavy metal, its inertia helps hold together the plutonium during the explosion in order to prevent the premature disintegration of the fissioning material and thereby obtain a greater efficiency.

In a nuclear explosion exceedingly high temperatures (hundreds of millions of degrees centigrade) and exceedingly high pressures (millions of atmospheres) build up very rapidly (in about one-half of a millionth of a second, the time taken for about 55 generations of fission). The mass of the material used for fission expands at very high speeds – initially at a speed of about 1,000 kilometres a second. In much less than a millionth of a second the size and density of the material have changed so that the mass becomes less than critical and the chain reaction stops. The designer of a nuclear weapon aims at keeping the fissionable material together, against its tendency to fly apart, long enough to produce an explosion powerful enough for his purpose.

The atomic bomb used to destroy Nagasaki in 1945 used about 8 kilograms of plutonium, containing more than 90 per cent of the isotope plutonium-239, surrounded by a tamper. The complete detonation of 1 kilogram of plutonium would produce an explosion equivalent to that of 18,000 tons (18 kilotons) of TNT. The 8 kilograms of plutonium used in the Nagasaki bomb produced an explosion equivalent to that of 22 kilotons of TNT. Its efficiency was therefore only about 15 per cent. Modern fission bombs have much greater efficiencies – approaching 40 per cent.

For maximum efficiency, the chain reaction in an atomic bomb must be initiated at precisely the right moment. The initiation is achieved by a pulse of neutrons. In the Nagasaki bomb, the initiator consisted of a hollow sphere placed at the centre of the plutonium core (which was in the form of two hemispheres so that the initiator could be assembled at the centre). Inside the initiator were some polonium and beryllium, two elements which produce neutrons when intimately mixed. The two substances were placed on opposite sides of the hollow sphere.

At the moment of implosion, the shock wave from the chemical explosion crushed the initiator, mixing the polonium and beryllium and producing a pulse of neutrons which initiated the chain reaction in the plutonium when the mass of plutonium was super-critical. In today's nuclear weapons, the neutron pulse is produced by a small electronic device called a neutron 'gun'.

A major problem in designing this type of nuclear weapon for maximum efficiency is to prevent the chain reaction from being started before the maximum achievable super-criticality is reached – an eventuality called pre-detonation. Pre-detonation is most likely to be caused by a neutron from spontaneous fission – fission that occurs naturally without the stimulation of an external neutron – in the material used for fission. In 8 kilograms of plutonium-239, for example, the average time between spontaneous fissions is only about three-millionths of a second. To prevent pre-detonation and loss of efficiency, the assembly of a plutonium bomb must be very rapid. Implosion is necessary.

An excellent description of a plutonium fission nuclear weapon is given by Margaret Gowing, in *Independence and Deterrence: Britain and Atomic Energy, 1945–52*. She describes an early British design (which she calls 'the gadget') as follows.

> An implosion design has been chosen, in which the mass of high explosive, surrounding a sphere containing both the fissile material and a tamper, was so arranged as to produce a shock wave travelling radially inwards and thus compressing the material. [Note: The high explosive was arranged in a number of shaped charges, called 'lenses'.] The design had the advantage of high velocities, which reduced the chance of pre-detonation despite the many background neutrons present in plutonium; at the same

time the material was compressed to such density that super-critical masses were obtained with comparatively little material. It had been realised at Los Alamos (the Manhattan Project) that performance could be improved by using explosive lenses to turn the divergent waves, which started from detonators, into parts of a common spherical wave converging on the centre of the sphere.

The main components of the gadget can be listed, working from outside to the centre. First came the detonators, which operated from an impulse from a firing device and involved other auxiliaries like safety switches and arming circuits. The detonation had to be started simultaneously in all the lenses; the lenses themselves were carefully calculated shapes, containing a combination of fast and slow explosive so that transit from the detonator to every point on the inner spherical surface of the lens was simultaneous. The detonation from the lenses then reached a spherical shell of homogenous high explosive called the supercharge. Within the supercharge was the tamper, which converted the divergent detonation wave into a convergent shock wave, reflected some of the neutrons back into the fissile material and generally increased the efficiency of the explosion. Within the tamper was the plutonium and within that the initiator. The last component was necessary because, although the implosion resulted in a powerful compression of the fissile material and the surrounding tamper the material would stay compressed only for a few microseconds and would then expand again very quickly. It was therefore essential to make sure that the chain started at the right moment. This could be done by creating at the centre of the fissile material an intense neutron source (Gowing, 1974).

The timing of the detonations of the chemical explosives is crucial for the efficient operation of an implosion atomic bomb. Micro-second (a millionth of a second) precision is essential. The shapes of the explosive lenses are rather complex and must be carefully calculated. The high explosive must be chemically extremely pure and of constant constituency throughout its volume. Getting the electronic timing and the chemistry and geometry of the explosive lenses right are the most difficult problems in designing an implosion-type atomic bomb. But competent electronic and explosive engineers, given adequate resources and access to the literature, could solve them without too much difficulty. Incidentally, in today's world, explosive lenses and detonators adequate for an implosion-type atomic bomb are commercially available.

The difficulty of designing and fabricating a nuclear weapon using fission is often exaggerated. A competent group of nuclear physicists and engineers would have no difficulty in designing from scratch and constructing a fission nuclear weapon. They would not need access to any classified literature. All the information they would require is in the open literature.

Plutonium from nuclear-power reactors

The current nuclear-weapon powers, including Israel, have built special military reactors to produce plutonium for their nuclear weapons. Plutonium is, however, produced as an inevitable by-product in all nuclear reactors. But the isotopic composition of the plutonium produced in reactors operated for different purposes varies. The plutonium specifically for military purposes is rich in the isotope plutonium-239. 'Weapon-grade' plutonium typically contains more than 93 per cent of the isotope plutonium-239 and is the material which produces the most efficient atomic bombs.

Plutonium produced in nuclear reactors operated to produce electricity in the most economical way typically contains only about 60 per cent plutonium-239. About 25 per cent is plutonium-240 (in weapon-grade plutonium the amount is typically about 7 per cent) and about 10 per cent is plutonium-241. Can this reactor-grade plutonium be used to produce nuclear explosions? This is an important question because, if it can, countries operating nuclear-power reactors for peaceful purposes, particularly electricity production, have access to plutonium that could be used to produce nuclear weapons. And, as the quantity of reactor-grade plutonium in the world increases, it becomes easier for sub-national groups to steal it and produce nuclear explosives. Reactor-grade plutonium is of interest here because some Middle East countries plan to operate nuclear-power reactors.

The critical mass of reactor-grade plutonium is somewhat higher than that of weapon-grade plutonium. The critical mass of typical reactor-grade plutonium in the form of a bare metal sphere surrounded by a natural uranium reflector, about 10 centimetres thick, is about 7 kilograms. Reactor-grade plutonium is more likely to be available as plutonium oxide rather than plutonium metal, and its critical mass is about 35 kilograms, if in spherical shape, with a radius of about 9 centimetres. Reactor-grade plutonium oxide in uncompacted powder form has a critical mass of about 875 kilograms if in a sphere, with a radius of about 45 centimetres.

That reactor-grade plutonium can be used to produce a nuclear weapon has been shown in the United States where at least two such devices have been built and tested. Nevertheless, it is widely believed that it is very unlikely to be used by a sub-national group to build a nuclear explosive device.

Amory Lovins, in an article in the scientific journal *Nature* entitled 'Nuclear weapons and power-reactor plutonium', explains that this belief is based on the following assumptions:

*that reactor-grade plutonium is far more hazardous to people dealing with it than weapon-grade plutonium;

*that a nuclear explosive device made from reactor-grade plutonium is much more likely to explode unintentionally;

*that such a device, if it explodes at all, will not explode violently enough to do much damage, nor to accomplish the main aims of the makers; and

*that its explosive yield is too unpredictable to be acceptable to its makers.

Lovins concludes that 'each of these assumptions contains, in certain circumstances, an element of truth', but he adds, 'each is generally, or can by plausible counter-measures be rendered, false. Their implication that reactor-grade plutonium is not very dangerous is wishful thinking, and causes the proliferation risks of civil nuclear activities to be gravely underestimated.'

That plutonium would present a sub-national group with serious handling problems cannot be denied. The material is very toxic and reactor-grade plutonium is several times more radioactive than weapon-grade plutonium. Nevertheless, as the Office of Technology Assessment of the US Congress points out in its publication *Nuclear Proliferation and Safeguards*, radioactivity and handling problems are manageable for both reactor-grade and weapon-grade plutonium. Lovins is more precise. He shows that total radiation dose rates from reactor-grade plutonium – including plutonium oxide – are such that they do not provide an effective deterrent. Given sensible precautions against achieving criticality accidentally – and thereby producing a very large burst of neutrons – a sub-national group constructing a nuclear explosive from reactor-grade plutonium would not face serious radiological hazards.

The main problem with using reactor-grade plutonium in nuclear weapons is that the spontaneous fission rate of plutonium-240 is much greater than that of plutonium-239. In reactor-grade plutonium the average time between spontaneous fissions is less than a micro-second (a millionth of a second). This means that very fast implosion techniques would be necessary in a nuclear device made from reactor-grade plutonium to prevent pre-detonation. Pre-detonation leads to uncertain explosive yields. The military would prefer to know rather precisely what the yields of their nuclear weapons are likely to be and would be unwilling to tolerate uncertain yields; but a sub-national group would probably not be particular about explosive yields.

Spontaneous fission produces a neutron background in a weapon-grade plutonium core of about one neutron every two or three micro-seconds. With a mean time of a few micro-seconds between neutrons – a very much longer time than the duration of the fission chain reaction – radial compression rates of a few millimetres per micro-second will prevent pre-detonation. Implosion techniques can achieve this without much difficulty. But for reactor-grade plutonium the mean time between neutrons is a small fraction of a micro-second. Extremely fast assembly would be needed to achieve super-criticality. Implosion technologies to provide the very high shock velocities and compression needed to prevent pre-detonation are available but probably not to a sub-national group, at least in the foreseeable future.

Pre-detonation does not make a nuclear explosive device unreliable. As the Office of Technology Assessment explains, pre-detonation produces

> a statistical uncertainty in the yield. Another way to state this is that the probable nuclear yield is statistically distributed between predictable upper and lower limits, which are likely to be more than a factor of ten apart. For a *well-understood design properly constructed*, however, the most probable yield range could be predicted within much closer limits (OTA, 1977).

Unpredictability of yield is unlikely seriously to bother a sub-national group intent on manufacturing a nuclear explosive. In fact, the plan of the group might well be to build a nuclear explosive, place it in the centre of a city, send proof to the authorities that they had acquired plutonium and then blackmail them. In this way, the group could achieve its ends, or believe that it could do so, without actually exploding its device. The device need, therefore, be only a very crude one.

Reprocessing

Plutonium-239 is produced in all nuclear reactors when uranium-238 nuclei in the reactor fuel absorb slow neutrons. In a reactor fuel element, the amount of plutonium increases as the uranium burns up. The plutonium can be removed from the reactor fuel element by a relatively straightforward chemical process.

If done commercially, for profit, a reprocessing plant is a

complex and costly chemical establishment and, because the capital cost is relatively independent of the size of the plant, economic reprocessing can only be achieved if a large-scale plant is used to serve many reactors. But to obtain plutonium for military purposes, where money is not the object, reprocessing can easily be done on a small scale.

In fact, plutonium can be obtained clandestinely from a nuclear reactor, acquired especially for the purpose. The components for a small reactor, capable of producing enough plutonium to make a few nuclear weapons a year, can be easily, and secretly, obtained on the open market for roughly $30 million (about the same cost as a modern fighter aircraft).

The reactor and a small reprocessing facility to remove the plutonium from the reactor fuel elements could be clandestinely constructed and operated. These units, and room to design and construct nuclear weapons from the plutonium, could be effectively disguised and hidden in a building or underground.

The use of enriched uranium in nuclear weapons

Natural uranium contains the isotopes uranium-235 and uranium-238. As dug out of the ground, uranium is mostly uranium-238 – out of a thousand atoms of natural uranium, only 7 are uranium-235. The problem with uranium-238 is that a neutron can only cause fission in it if its velocity exceeds a certain value. But too few of the neutrons available from the fission process have more than this critical velocity. A nucleus of uranium-235, on the other hand, like a nucleus of plutonium-239, will undergo fission when even a slow-moving neutron collides with it, and so a chain reaction is possible using uranium-235.

To obtain uranium that can be used to construct an atomic bomb, the amount of uranium-235 in natural uranium is increased by a process called enrichment. The proportion of uranium-235 is normally enriched from its natural value of 0.7 per cent to more than 40 per cent, preferably to over 95 per cent. The greater the amount of uranium-235 in the uranium, the less will be the critical mass.

Three methods are available for enriching uranium – the gaseous diffusion, the gas centrifuge and the jet nozzle techniques. The technique mainly used so far is the gas diffusion method. But, given the materials now available, particularly carbon fibre, countries wishing to enrich uranium would probably opt for the gas centrifuge method.

A gas centrifuge for uranium enrichment consists of a vacuum tank containing a long, rotating drum with a nozzle at one end and an orifice at the other. Uranium hexafluoride gas is pumped in via the nozzle and, as it moves up inside the rotating drum, molecules of it will tend to be flung outwards by the centrifugal force. Molecules of uranium hexafluoride gas in which the uranium is uranium-238 are slightly heavier than those of uranium-235 hexafluoride. There will, therefore, be a difference in the centrifugal force acting on the molecules of different masses when the gas is rotated at very high speed. Molecules of the lighter uranium-235 isotope will diffuse towards the centre, the inner portion thus becomes enriched in uranium-235 and this is collected at the exit orifice.

A plant containing many gas centrifuges in a cascade is needed to enrich a useful quantity of uranium. The slightly enriched flow of uranium gas from the first centrifuge is fed into the nozzle of the next centrifuge in the cascade and so on. The uranium is circulated around the cascade until the desired degree of enrichment is obtained.

In uranium-235, the average time between spontaneous fissions is much greater than it is in plutonium-239 and the so-called 'gun' method can be used to assemble a critical mass of uranium-235 in an atomic bomb. In the Hiroshima bomb, for example, a less than critical mass of uranium-235 was fired down a 'cannon barrel' into another less than critical mass of uranium-235 placed in front of the 'muzzle'. When the two masses came together they formed a super-critical mass which exploded. About 60 kilograms of uranium-235 were used in the Hiroshima bomb, and about 700 grams were fissioned. The average time between spontaneous fissions was about one-fiftieth of a second – quite adequate for the gun technique. The yield of the Hiroshima bomb was equivalent to that of about 12.5 kilotons of TNT.

If surrounded by a reflector made from natural uranium 15 centimetres thick, 100 per cent pure uranium-235 has a critical mass of 15 kilograms (compared with 4.4 kilograms for plutonium-239). With uranium enriched to 40 per cent uranium-235, the critical mass increases to 75 kilograms; with 20 per cent uranium-235, it is 250 kilograms. High concentrations of uranium-235 are, therefore, highly desirable if the material is to be used to produce nuclear weapons.

Designs based on the Hiroshima and Nagasaki bombs are likely to be used by countries beginning a nuclear-weapon programme, but even the first weapons now produced would probably be more sophisticated than these early primitive weapons. The Nagasaki bomb, for example, was about 3 metres long, 1.5 metres wide, and weighed about 4.5 tons. A modern fission weapon, even the first produced in a nuclear-weapon programme, should weigh no more than a few hundred kilograms.

A small plant containing gas centrifuges to enrich uranium for a nuclear-weapon programme could, like a military plutonium facility, be constructed clandestinely. Because the basic design of nuclear weapons using nuclear fission is now well known and the nuclear data needed are in the open literature, and because the facilities to produce plutonium and enriched uranium for nuclear weapons can be built and operated secretly and simply, we do not know for sure which countries have atomic bombs and which do not.

H-BOMBS

There is a limit to the explosive power that can be obtained from an operational nuclear weapon based on pure fission, the maximum being equivalent to the explosion of a few tens of kilotons of TNT. If larger explosions are required extra energy must be obtained from fusion.

The fusion process is the opposite of fission. In fusion, light nuclei are formed (i.e. fused) into heavier nuclei. In nuclear weapons, the heavier isotopes of hydrogen – deuterium and tritium – are fused together to form helium. The fusion process, like the fission process, produces energy and is accompanied by the emission of neutrons. There is no critical mass for the fusion process and therefore, in theory, there is no limit to the explosive yield of fusion weapons – or H-bombs (H for hydrogen) as they are often called.

Fission is relatively easy to initiate; one neutron will start a chain reaction going in a critical mass of a fissionable material, such as plutonium-239 or uranium-235. But fusion is possible only if the nuclei to be fused together are given a high enough energy to overcome the repulsive electric force between them due to their positive electric charges. In H-bombs, this energy is provided by raising the temperature of the fusion material. Because H-bombs depend on heat they are also called thermonuclear weapons.

In a typical thermonuclear weapon, deuterium and tritium are fused together. But to get this fusion reaction to work, the deuterium-tritium mixture must be raised to a temperature of a hundred million degrees Centigrade or so. This can be provided only by an atomic bomb in which such a temperature occurs at the moment of the explosion. An H-bomb, therefore, consists of a fission stage, which is an atomic bomb acting as a trigger, and a fusion stage, in which hydrogen isotopes (tritium and deuterium) are fused by the heat produced by the trigger.

The simplest fusion bomb is the so-called 'boosted weapon'. In a boosted weapon, some fusion material is placed at the centre of the plutonium sphere in an ordinary atomic bomb. Fusion energy produced by the explosion of the atomic bomb adds to the fission energy and increases the explosive power of the weapon. Moreover, the neutrons produced during the fusion process produce additional fissions in the plutonium in the weapon, increasing its efficiency even more.

Actually, in boosted weapons of this type, fusion is used mainly as a source of neutrons to help the fission process, rather than as a direct source of energy. Boosted weapons are essentially sophisticated fission weapons. A much higher explosive power is obtained from a given amount of plutonium.

It is impracticable to produce pure fission weapons, with no boosting, that will have explosive yields greater than that of 50,000 tons (50 kilotons) or so of TNT. Militarily usable boosted weapons, however, have explosive powers equivalent to that of 100 to 200 kilotons of TNT. If still higher explosive yields are required a 'staged' thermonuclear weapon must be used.

In a staged device, the fusion material is placed outside the plutonium sphere, forming a second stage that is triggered by the fission explosion. Normally, the fusion material is in the form of cylinder, which is made out of lithium deuteride. When neutrons from the fission explosion bombard lithium nuclei in the lithium deuteride, tritium nuclei are produced, and these fuse with deuterium nuclei in the lithium deuteride to produce fusion energy.

It is very advantageous to use lithium deuteride as the fusion material because it is a solid at normal temperatures, whereas tritium and deuterium, the fusion materials used in boosted weapons, are gases at normal temperatures. It is, of course, much easier to construct nuclear weapons from solid materials than gases.

The energy released from such a thermonuclear weapon comes from the fission trigger and the fusion material. But, if the fusion device is surrounded by a shell of uranium metal, the high-energy neutrons produced in the fusion process will cause additional fissions in the uranium shell. This technique can be used to enhance considerably the explosive power of a thermonuclear weapon. Such a weapon is called a fission-fusion-fission device. On average, about half of the yield from a typical thermonuclear weapon will come from fission and the other half from fusion.

H-bombs are much more difficult to design than A-bombs. The problem is to prevent the A-bomb trigger from blowing the whole weapon apart before enough fusion material has been ignited to give the required explosive yield. Sufficient energy has to be delivered to the fusion material to start the thermonuclear reaction in a time much shorter than the time it takes for the explosion to occur. This means that the energy must be delivered with a speed approaching the speed of light.

Professor J. Rotblat has described the technique used:

> The solution to the problem lies in the fact that at the very high temperature of the fission trigger most of the energy is emitted in the form of X-rays. These X-rays, travelling with the speed of light, radiate out from the centre and on reaching the tamper (surrounding the fusion material) are absorbed in it and then immediately re-emitted in the form of softer X-rays. By an appropriate configuration of the trigger and the fusion material it is possible to ensure that the X-rays reach the latter almost instantaneously. If the fusion material is sub-divided into small portions, each surrounded with a thin absorber made of a heavy metal, the bulk of the fusion material will simultaneously receive enough energy to start the thermonuclear reaction before the explosion disperses the whole assembly (Rotblat, 1981).

Although essentially weightless, X-rays can exert great pressure. In an H-bomb, the pressure (several million pounds per square inch) is exerted uniformly on the fusion material and long enough for the fusion process to work before the material is blown apart. Because the radiation travels at the speed of light, it arrives at the fusion material about a millionth of a second before the much slower moving shock wave from the trigger explosion. When the shock wave arrives, and blows the assembly apart, the fusion explosion has already occurred.

Very large explosive yields have been obtained with thermonuclear weapons. For example, the Soviet Union exploded an H-bomb in 1962 with a yield equal to that of 58 million tons of TNT – equivalent to about 3,000 Nagasaki bombs. Even higher yields could be obtained.

In summary, a country wishing to make ordinary fission weapons needs a supply of plutonium or enriched uranium.

Plutonium is made in nuclear reactors. Uranium is most easily enriched in gas centrifuges.

Any country capable of operating a nuclear reactor and/or constructing a gas centrifuge plant can manufacture ordinary fission weapons. In fact, a competent sub-national group could design and build a nuclear explosive device, even from plutonium produced in nuclear-power reactors run for normal electricity production.

The explosive yields of militarily useful fission nuclear weapons is limited. If more powerful nuclear weapons are required, they must be boosted by some fusion energy. A country wishing to make boosted nuclear weapons must have access to a supply of tritium and deuterium. For even higher explosive yields, full-scale thermonuclear weapons must be used. For these, lithium deuteride is necessary.

The production of a thermonuclear force is much more difficult than that of a fission nuclear force. Nevertheless, China managed to set off a full-scale thermonuclear explosion within two and a half years of its first fission nuclear test; for the USSR, the time lag was four years; for the UK, it was four and a half years; for France, it was eight and a half years; and for the USA, it was seven years. Israel's nuclear scientists are competent enough to produce the thermonuclear weapons within a few years of producing fission nuclear weapons.

Appendix II

TREATY ON THE NON-PROLIFERATION OF NUCLEAR WEAPONS

Signed at London, Moscow and Washington on 1 July 1968.
Entered into force on 5 March 1970.
Depositaries: UK, US and Soviet governments.

The States concluding this Treaty, hereinafter referred to as the 'Parties to the Treaty',

Considering the devastation that would be visited upon all mankind by a nuclear war and the consequent need to make every effort to avert the danger of such a war and to take measures to safeguard the security of peoples,

Believing that the proliferation of nuclear weapons would seriously enhance the danger of nuclear war,

In conformity with resolutions of the United Nations General Assembly calling for the conclusion of an agreement on the prevention of wider dissemination of nuclear weapons,

Undertaking to cooperate in facilitating the application of International Atomic Energy Agency safeguards on peaceful nuclear activities,

Expressing their support for research, development and other efforts to further the application, within the framework of the International Atomic Energy Agency safeguards system, of the principle of safeguarding effectively the flow of source and special fissionable materials by use of instruments and other techniques at certain strategic points,

Affirming the principle that the benefits of peaceful applications of nuclear technology, including any technological by-products which may be derived by nuclear-weapon States from the development of nuclear explosive devices, should be available for peaceful purposes to all Parties to the Treaty, whether nuclear-weapon or non-nuclear-weapon States,

Convinced that in furtherance of this principle, all Parties in the Treaty are entitled to participate in the fullest possible exchange of scientific information for, and to contribute alone or in cooperation with other States to, the further development of the applications of atomic energy for peaceful purposes,

Declaring their intention to achieve at the earliest possible date the cessation of the nuclear arms race and to undertake effective measures in the direction of nuclear disarmament,

Urging the cooperation of all States in the attainment of this objective,

Recalling the determination expressed by the Parties to the 1963 Treaty banning nuclear weapon tests in the atmosphere, in outer space and under water in its Preamble to seek to achieve the discontinuance of all test explosions of nuclear weapons for all time and to continue negotiations to this end,

Desiring to further the easing of international tension and the strengthening of trust between States in order to facilitate the cessation of the manufacture of nuclear weapons, the liquidation of all their existing stockpiles, and the elimination from national arsenals of nuclear weapons and the means of their delivery pursuant to a treaty on general and complete disarmament under strict and effective international control,

Recalling that, in accordance with the Charter of the United Nations, States must refrain in their international relations from the threat of use of force against the territorial integrity or political independence of any State, or in any other manner inconsistent with the Purposes of the United Nations, and that the establishment and maintenance of international peace and security are to be promoted with the least diversion for armaments of the world's human and economic resources,

Have agreed as follows:

Article I

Each nuclear-weapon State Party to the Treaty undertakes not to transfer to any recipient whatsoever nuclear weapons or other nuclear explosive devices or control over such weapons or explosive devices directly, or indirectly; and not in any way to assist, encourage, or induce any non-nuclear-weapon State to manufacture or otherwise acquire nuclear weapons or other nuclear explosive devices, or control over such weapons or explosive devices.

Article II

Each non-nuclear-weapon State Party to the Treaty undertakes not to receive the transfer from any transferor whatsoever of nuclear weapons or other nuclear explosive devices or of control over such weapons or explosive devices directly, or indirectly; not to manufacture or otherwise acquire nuclear weapons or other nuclear explosive devices; and not to seek or receive any assistance in the manufacture of nuclear weapons or other nuclear explosive devices.

Article III

1. Each non-nuclear-weapon State Party to the Treaty undertakes to accept safeguards, as set forth in an agreement to be negotiated and concluded with the International Atomic Energy Agency in accordance with the Statute of the International Atomic Energy Agency and the Agency's safeguards system, for the exclusive purpose of verification of the fulfillment of its obligations assumed under this Treaty with a view to preventing diversion of nuclear energy from peaceful uses to nuclear weapons or other nuclear explosive devices. Procedures for the safeguards required by this article shall be followed with respect to source or special fissionable material whether it is being produced, processed or used in any principal nuclear facility or is outside any such facility. The safeguards required by this article shall be applied on all source or special fissionable material in all peaceful nuclear activities within the territory of such State, under its jurisdiction, or carried out under its control anywhere.

2. Each State Party to the Treaty undertakes not to provide: (a) source or special fissionable material, or (b) equipment or material especially designed or prepared for the processing, use or production of special fissionable material, to any non-nuclear-weapon State for peaceful purposes, unless the source or special fissionable material shall be subject to the safeguards required by this article.

3 The safeguards required by this article shall be implemented in a manner designed to comply with article IV of this Treaty, and to avoid hampering the economic or technological development of the Parties or international cooperation in the field of peaceful nuclear activities, including the international exchange of nuclear material and equipment for the processing, use or production of nuclear material for peaceful purposes in accordance with the provisions of this article and the principle of safeguarding set forth in the Preamble of the Treaty.

4. Non-nuclear-weapon States Party to the Treaty shall conclude agreements with the International Atomic Energy Agency to meet the requirements of this article either individually or together with other States in accordance with the Statue of the International Atomic Energy Agency. Negotiation of such agreements shall commence within 180 days from the original entry into force of this Treaty. For States depositing their instruments of ratification or accession after the 180-day period, negotiation of such agreements shall commence not later than the date of such deposit. Such agreements shall enter into force not later than eighteen months after the date of initiation of negotiations.

Article IV

1. Nothing in this Treaty shall be interpreted as affecting the inalienable right of all the Parties to the Treaty to develop research, production and use of nuclear energy for peaceful purposes without discrimination and in conformity with articles I and II of this Treaty.

2. All the Parties to the Treaty undertake to facilitate, and have the right to participate in, the fullest possible exchange of equipment, materials and scientific and technological information for the peaceful uses of nuclear energy. Parties to the Treaty in a position to do so shall also cooperate in contributing alone or together with other States or international organizations to the further development of the applications of nuclear energy for peaceful purposes, especially in the territories of non-nuclear-weapon States Party to the Treaty, with due consideration for the needs of the developing areas of the world.

Article V

Each Party to the Treaty undertakes to take appropriate measures to ensure that, in accordance with this Treaty, under appropriate international observation and through appropriate international procedures, potential benefits from any peaceful applications of nuclear explosions will be made available to non-nuclear-weapon States Party to the Treaty on a non-discriminatory basis and that the charge to such Parties for the explosive devices used will be as low as possible and exclude any charge for research and development. Non-nuclear-weapon States Party to the Treaty shall be able to obtain such benefits, pursuant to a special international agreement or agreements, through an appropriate international body with adequate representation of non-nuclear-weapon States. Negotiations on this subject shall commence as soon as possible after the Treaty enters into force. Non-nuclear-weapon States Party to the Treaty so desiring may also obtain such benefits pursuant to bilateral agreements.

Article VI

Each of the Parties to the Treaty undertakes to pursue negotiations in good faith on effective measures relating to cessation of the nuclear arms race at an early date and to nuclear disarmament, and on a treaty on general and complete disarmament under strict and effective international control.

Article VII

Nothing in this Treaty affects the right of any group of States to conclude regional treaties in order to assure the total absence of nuclear weapons in their respective territories.

Article VIII

1. Any Party to the Treaty may propose amendments to this Treaty. The text of any proposed amendment shall be submitted to the Depositary Governments which shall circulate it to all Parties to the Treaty. Thereupon, if requested to do so by one-third or more of the Parties to the Treaty, the Depositary Governments shall convene a conference, to which they shall invite all the Parties to the Treaty, to consider such an amendment.

2. Any amendment to this Treaty must be approved by a majority of the votes of all the Parties to the Treaty, including the votes of all nuclear-weapon States Party to the Treaty and all other Parties which, on the date the amendment is circulated, are members of the Board of Governors of the International Atomic Energy Agency. The amendment shall enter into force for each Party that deposits its instrument of ratification of the amendment upon the deposit of such instruments of ratification by a majority of all the Parties, including the instruments of ratification of all nuclear-weapon States Party to the Treaty and all other Parties which, on the date the amendment is circulated, are members of the Board of Governors of the International Atomic Energy Agency. Thereafter, it shall enter into force for any other Party upon the deposit of its instrument of ratification of the amendment.

3. Five years after the entry into force of this Treaty, a conference of Parties to the Treaty shall be held in Geneva, Switzerland, in order to review the operation of this Treaty with a

view to assuring that the purposes of the Preamble and the provisions of the Treaty are being realized. At intervals of five years thereafter, a majority of the Parties to the Treaty may obtain, by submitting a proposal to this effect to the Depositary Governments, the convening of further conferences with the same objective of reviewing the operation of the Treaty.

Article IX

1. This Treaty shall be open to all States for signature. Any State which does not sign the Treaty before its entry into force in accordance with paragraph 3 of this article may accede to it at any time.

2. This Treaty shall be subject to ratification by signatory States. Instruments of ratification and instruments of accession shall be deposited with the Governments of the United States of America, the United Kingdom of Great Britain and Northern Ireland and the Union of Soviet Socialist Republics, which are hereby designated the Depositary Governments.

3. This Treaty shall enter into force after its ratification by the States, the Governments of which are designated Depositaries of the Treaty, and forty other States signatory to this Treaty and the deposit of their instruments of ratification. For the purposes of this Treaty, a nuclear-weapon State is one which has manufactured and exploded a nuclear weapon or other nuclear explosive device prior to January 1, 1967.

4. For States whose instruments of ratification or accession are deposited subsequent to the entry into force of this Treaty, it shall enter into force on the date of the deposit of their instruments of ratification or accession.

5. The Depositary Governments shall promptly inform all signatory and acceding States of the date of each signature, the date of deposit of each instrument of ratification or of accession, the date of the entry into force of this Treaty, and the date of receipt of any requests for convening a conference or other notices.

6. This Treaty shall be registered by the Depositary Governments pursuant to article 102 of the Charter of the United Nations.

Article X

1. Each Party shall in exercising its national sovereignty have the right to withdraw from the Treaty if it decides that extraordinary events, related to the subject matter of this Treaty,

have jeopardized the supreme interests of its country. It shall give notice of such withdrawal to all other Parties to the Treaty and to the United Nations Security Council three months in advance. Such notice shall include a statement of the extraordinary events it regards as having jeopardized its supreme interests.

2. Twenty-five years after the entry into force of the Treaty, a conference shall be convened to decide whether the Treaty shall continue in force indefinitely, or shall be extended for an additional fixed period or periods. This decision shall be taken by a majority of the Parties to the Treaty.

Article XI

This Treaty, the English, Russian, French, Spanish and Chinese texts of which are equally authentic, shall be deposited in the archives of the Depositary Governments. Duly certified copies of this Treaty shall be transmitted by the Depositary Governments to the Governments of the signatory and acceding States.

Source: *United States Treaties and Other International Agreements*, Vol. 21, Part 1 (US Department of State, Washington, 1970).

TREATY BANNING NUCLEAR WEAPON TESTS IN THE ATMOSPHERE, IN OUTER SPACE AND UNDER WATER

Signed at Moscow on 5 August 1963.
Entered into force on 10 October 1963.
Depositaries: UK, US and Soviet governments.

The Governments of the United States of America, the United Kingdom of Great Britain and Northern Ireland, and the Union of Soviet Socialist Republics, hereinafter referred to as the 'Original Parties',

Proclaiming as their principal aim the speediest possible achievement of an agreement on general and complete disarmament under strict international control in accordance with the objectives of the United Nations which would put an end to the armaments race and eliminate the incentive to the production and testing of all kinds of weapons, including nuclear weapons,

Seeking to achieve the discontinuance of all test explosions of nuclear weapons for all time, determined to continue negotiations to this end, and desiring to put an end to the contamination of man's environment by radioactive substances,

Have agreed as follows:

Article I

1. Each of the Parties to this Treaty undertakes to prohibit, to prevent, and not to carry out any nuclear weapon test explosion, or any other nuclear explosion, at any place under its jurisdiction or control:

(*a*) in the atmosphere; beyond its limits, including outer space; or under water, including territorial waters or high seas; or

(*b*) in any other environment if such explosion causes radioactive debris to be present outside the territorial limits of the State under whose jurisdiction or control such explosion is conducted. It is understood in this connection that the provisions of this subparagraph are without prejudice to the conclusion of a treaty resulting in the permanent banning of all nuclear test explosions, including all such explosions underground, the conclusion of which, as the Parties have stated in the Preamble to this Treaty, they seek to achieve.

2. Each of the Parties to this Treaty undertakes furthermore to refrain from causing, encouraging, or in any way participating in, the carrying out of any nuclear weapon test explosion, or any other nuclear explosion, anywhere which would take place in any of the environments described, or have the effect referred to, in paragraph 1 of this Article.

Article II

1. Any Party may propose amendments to this Treaty. The text of any proposed amendment shall be submitted to the Depositary Governments which shall circulate it to all Parties to this Treaty. Thereafter, if requested to do so by one-third or more of the Parties, the Depositary Governments shall convene a conference, to which they shall invite all the Parties, to consider such amendment.

2. Any amendment to this Treaty must be approved by a majority of the votes of all the Parties to this Treaty, including the votes of all of the Original Parties. The amendment shall enter into force for all Parties upon the deposit of instruments of ratification by a majority of all the Parties, including the instruments of ratification of all of the Original Parties.

Article III

1. This Treaty shall be open to all States for signature. Any State which does not sign this Treaty before its entry into force in

accordance with paragraph 3 of this Article may accede to it at any time.

2. This Treaty shall be subject to ratification by signatory States. Instruments of ratification and instruments of accession shall be deposited with the Governments of the Original Parties – the United States of America, the United Kingdom of Great Britain and Northern Ireland, and the Union of Soviet Socialist Republics – which are hereby designated the Depositary Governments.

3. This Treaty shall enter into force after its ratification by all the Original Parties and the deposit of their instruments of ratification.

4. For States whose instruments of ratification or accession are deposited subsequent to the entry into force of this Treaty, it shall enter into force on the date of the deposit of their instruments of ratification or accession.

5. The Depositary Governments shall promptly inform all signatory and acceding States of the date of each signature, the date of deposit of each instrument of ratification of and accession to this Treaty, the date of its entry into force, and the date of receipt of any requests for conferences or other notices.

6. This Treaty shall be registered by the Depositary Governments pursuant to Article 102 of the Charter of the United Nations.

Article IV

This Treaty shall be of unlimited duration.

Each Party shall in exercising its national sovereignty have the right to withdraw from the Treaty if it decides that extraordinary events, related to the subject matter of this Treaty, have jeopardized the supreme interests of its country. It shall give notice of such withdrawal to all other Parties to the Treaty three months in advance.

Article V

This Treaty, of which the English and Russian texts are equally authentic, shall be deposited in the archives of the Depositary Governments. Duly certified copies of this Treaty shall be transmitted by the Depositary Governments to the Governments of the signatory and acceding States.

In witness whereof the undersigned, duly authorized, have signed this Treaty.

Done in triplicate at the city of Moscow the fifth day of August, one thousand nine hundred and sixty-three.

Source: *Treaty Series*, Col. 480 (United Nations, New York, 1963).

TREATY FOR THE PROHIBITION OF NUCLEAR WEAPONS IN LATIN AMERICA

Signed at Mexico, Federal District, on 14 February 1967.

The Treaty enters into force for each state that has ratified it when the requirements specified in the Treaty have been met, that is, that all states in the region which were in existence when the Treaty was opened for signature, deposit the instruments of ratification; that Protocols I and II be signed and ratified by those states to which they apply; and that agreements on safeguards be concluded with the IAEA. The signatory states have the right to waive, wholly or in part, those requirements.

The Treaty came into force on 22 April 1968 as between Mexico and El Salvador, on behalf of which instruments of ratification, with annexed declarations wholly waiving the above requirements, were deposited on 20 September 1967 and 22 April 1968, respectively.

The Protocols enter into force for the states that have ratified them on the date of the deposit of their instruments of ratification.

Depositary: Government of Mexico.

Preamble

In the name of their peoples and faithfully interpreting their desires and aspirations, the Governments of the States which sign the Treaty for the Prohibition of Nuclear Weapons in Latin America,

Desiring to contribute, so far as lies in their power, towards ending the armaments race, especially in the field of nuclear weapons, and towards strengthening a world at peace, based on the sovereign equality of States, mutual respect and good neighbourliness,

Recalling that the United Nations General Assembly, in its Resolution 808 (IX), adopted unanimously as one of the three points of a coordinated programme of disarmament 'the total prohibition of the use and manufacture of nuclear weapons and weapons of mass destruction of every type',

Recalling that militarily denuclearized zones are not an end in themselves but rather a means for achieving general and complete disarmament at a later stage,

Recalling United Nations General Assembly Resolution 1911 (XVIII), which established that the measures that should be agreed upon for the denuclearization of Latin American should be taken 'in the light of the principles of the Charter of the United Nations and of regional agreements',

Recalling United Nations General Assembly Resolution 2028 (XX), which established the principle of an acceptable balance of mutual responsibilities and duties for the nuclear and non-nuclear powers, and

Recalling that the Charter of the Organization of American States proclaims that it is an essential purpose of the Organization to strengthen the peace and security of the hemisphere,

Convinced:

That the incalculable destructive power of nuclear weapons has made it imperative that the legal prohibition of war should be strictly observed in practice if the survival of civilization and of mankind itself is to be assured,

That nuclear weapons, whose terrible effects are suffered, indiscriminately and inexorably, by military forces and civilian population alike, constitute, through the persistence of the radioactivity they release, an attack on the integrity of the human species and ultimately may even render the whole earth uninhabitable,

That general and complete disarmament under effective international control is a vital matter which all the peoples of the world equally demand,

That the proliferation of nuclear weapons, which seems inevitable unless States, in the exercise of their sovereign rights, impose restrictions on themselves in order to prevent it, would make any agreement on disarmament enormously difficult and would increase the danger of the outbreak of a nuclear conflagration,

That the establishment of militarily denuclearized zones is closely linked with the maintenance of peace and security in the respective regions,

That the military denuclearization of vast geographical zones, adopted by the sovereign decision of the States comprised therein, will exercise a beneficial influence on other regions where similar conditions exist,

That the privileged situation of the signatory States, whose territories are wholly free from nuclear weapons, imposes upon them the inescapable duty of preserving that situation both in their own interests and for the good of mankind,

That the existence of nuclear weapons in any country of Latin America would make it a target for possible nuclear attacks and would inevitably set off, throughout the region, a ruinous race in nuclear weapons which would involve the unjustifiable diversion, for warlike purposes, of the limited resources required for economic and social development,

That the foregoing reasons, together with the traditional peace-loving outlook of Latin America, give rise to an inescapable necessity that nuclear energy should be used in that region exclusively for peaceful purposes, and that the Latin American countries should use their right to the greatest and most equitable possible access to this new source of energy in order to expedite the economic and social development of their peoples,

Convinced finally:

That the military denuclearization of Latin America – being understood to mean the undertaking entered into internationally in this Treaty to keep their territories forever free from nuclear weapons – will constitute a measure which will spare their peoples from the squandering of their limited resources on nuclear armaments and will protect them against possible nuclear attacks on their territories, and will also constitute a significant contribution towards preventing the proliferation of nuclear-weapons and a powerful factor for general and complete disarmament, and

That Latin America, faithful to its tradition of universality, must not only endeavour to banish from its homelands the scourge of a nuclear war, but must also strive to promote the well-being and advancement of its peoples, at the same time co-operating in the fulfilment of the ideals of mankind, that is to say, in the consolidation of a permanent peace based on equal rights, economic fairness and social justice for all, in accordance with the principles and purposes set forth in the Charter of the United Nations and in the Charter of the Organization of American States,

Have agreed as follows:

Article I. *Obligations*

1. The Contracting Parties hereby undertake to use exclusively

for peaceful purposes the nuclear material and facilities which are under their jurisdiction, and to prohibit and prevent in their respective territories:

(*a*) The testing, use, manufacture, production or acquisition by any means whatsoever of any nuclear weapons, by the Parties themselves, directly or indirectly, on behalf of anyone else or in any other way, and

(*b*) The receipt, storage, installation, deployment and any form of possession of any nuclear weapons, directly or indirectly, by the Parties themselves, by anyone on their behalf or in any other way.

2. The Contracting Parties also undertake to refrain from engaging in, encouraging or authorizing, directly or indirectly, or in any way participating in the testing, use, manufacture, production, possession or control of any nuclear weapon.

Article 2. *Definition of the Contracting Parties*

For the purposes of this Treaty, the Contracting Parties are those for whom the Treaty is in force.

Article 3. *Definition of territory*

For the purposes of this Treaty, the term 'territory' shall include the territorial sea, air space and any other space over which the State exercises sovereignty in accordance with its own legislation.

Article 4. *Zone of application*

1. The zone of application of this Treaty is the whole of the territories for which the Treaty is in force.

2. Upon fulfilment of the requirements of article 28, paragraph 1, the zone of application of this Treaty shall also be that which is situated in the western hemisphere within the following limits (except the continental part of the territory of the United States of America and its territorial waters): starting at a point located at 35° north latitude, 75° west longitude; from this point directly southward to a point at 30° north latitude, 75° west longitude; from there, directly eastward to a point at 30° north latitude, 50° west longitude; from there, along a loxodromic line to a point at 5° north latitude, 20° west longitude; from there, directly southward to a point at 60° south latitude, 20° west longitude; from there, directly westward to a point at 60° south latitude, 115° west longitude; from there, directly northward to a point at 0 latitude, 115° west longitude; from there, along a loxodromic line to a point

at 35° north latitude, 150° west longitude; from there, directly eastward to a point at 35° north latitude, 75° west longitude.

Article 5. *Definition of nuclear weapons*

For the purposes of this Treaty, a nuclear weapon is any device which is capable of releasing nuclear energy in an uncontrolled manner and which has a group of characteristics that are appropriate for use for warlike purposes. An instrument that may be used for the transport or propulsion of the device is not included in this definition if it is separable from the device and not an indivisible part thereof.

Article 6. *Meeting of signatories*

At the request of any of the signatory States or if the Agency established by article 7 should so decide, a meeting of all the signatories may be convoked to consider in common questions which may affect the very essence of this instrument, including possible amendments to it. In either case, the meeting will be convoked by the General Secretary.

Article 7. *Organization*

1. In order to ensure compliance with the obligations of this Treaty, the Contracting Parties hereby establish an international organization to be known as the Agency for the Prohibition of Nuclear Weapons in Latin America, hereinafter referred to as 'the Agency'. Only the Contracting Parties shall be affected by its decisions.
2. The Agency shall be responsible for the holding of periodic or extraordinary consultations among Member States on matters relating to the purposes, measures and procedures set forth in this Treaty and to the supervision of compliance with the obligations arising therefrom.
3. The Contracting Parties agree to extend to the Agency full and prompt co-operation in accordance with the provisions of this Treaty, of any agreements they may conclude with the Agency and of any agreements the Agency may conclude with any other international organization or body.
4. The headquarters of the Agency shall be in Mexico City.

Article 9. *Organs*

1. There are hereby established as principal organs of the Agency a General Conference, a Council and a Secretariat.
2. Such subsidiary organs as are considered necessary by the

General Conference may be established within the purview of this Treaty.

Article 9. *The General Conference*

1. The General Conference, the supreme organ of the Agency, shall be composed of all the Contracting Parties; it shall hold regular sessions every two years, and may also hold special sessions whenever this Treaty so provides or, in the opinion of the Council, the circumstances so require.

2. The General Conference:

(*a*) May consider and decide on any matters or questions covered by this Treaty, within the limits thereof, including those referring to powers and functions of any organ provided for in this Treaty;

(*b*) Shall establish procedures for the control system to ensure observance of this Treaty in accordance with its provisions;

(*c*) Shall elect the Members of the Council and the General Secretary;

(*d*) May remove the General Secretary from office if the proper functioning of the Agency so requires;

(*e*) Shall receive and consider the biennial and special reports submitted by the Council and the General Secretary.

(*f*) Shall initiate and consider studies designed to facilitate the optimum fulfilment of the aims of this Treaty, without prejudice to the power of the General Secretary independently to carry out similar studies for submission to and consideration by the Conference.

(*g*) Shall be the organ competent to authorize the conclusion of agreements with Governments and other international organizations and bodies.

3. The General Conference shall adopt the Agency's budget and fix the scale of financial contributions to be paid by Member States, taking into account the systems and criteria used for the same purpose by the United Nations.

4. The General Conference shall elect its officers for each session and may establish such subsidiary organs as it deems necessary for the performance of its functions.

5. Each Member of the Agency shall have one vote. The decisions of the General Conference shall be taken by a two-thirds majority of the Members present and voting in the case of matters relating to the control system and measures referred to in article 20, the admission of new Members, the election or removal of the General Secretary, adoption of the budget and matters related

thereto. Decisions on other matters, as well as procedural questions and also determination of which questions must be decided by a two-thirds majority, shall be taken by a simple majority of the Members present and voting.

6. The General Conference shall adopt its own rules of procedure.

Article 10. *The Council*

1. The Council shall be composed of five Members of the Agency elected by the General Conference from among the Contracting Parties, due account being taken of equitable geographic distribution.

2. The Members of the Council shall be elected for a term of four years. However, in the first election three will be elected for two years. Outgoing Members may not be re-elected for the following period unless the limited number of States for which the Treaty is in force so requires.

3. Each member of the Council shall have one representative.

4. The Council shall be so organized as to be able to function continuously.

5. In addition to the functions conferred upon it by this Treaty and to those which may be assigned to it by the General Conference, the Council shall, through the General Secretary, ensure the proper operation of the control system in accordance with the provisions of this Treaty and with the decisions adopted by the General Conference.

6. The Council shall submit an annual report on its work to the General Conference as well as such special reports as it deems necessary or which the General Conference requests of it.

7. The Council shall elect its officer for each session.

8. The decisions of the Council shall be taken by a simple majority of its Members present and voting.

9. The Council shall adopt its own rules of procedure.

Article 11. *The Secretariat*

1. The Secretariat shall consist of a General Secretary, who shall be the chief administrative officer of the Agency, and of such staff as the Agency may require. The term of office of the General Secretary shall be four years and he may be re-elected for a single additional term. The General Secretary may not be a national of the country in which the Agency has its headquarters. In case the office of General Secretary becomes vacant, a new election shall be held to fill the office for the remainder of the term.

2. The staff of the Secretariat shall be appointed by the General Secretary, in accordance with rules laid down by the General Conference.

3. In addition to the functions conferred upon him by this Treaty and to those which may be assigned to him by the General Conference, – the General Secretary shall ensure, as provided by article 10, paragraph 5, the proper operation of the control system established by this Treaty, in accordance with the provisions of the Treaty and the decisions taken by the General Conference.

4. The General Secretary shall act in that capacity in all meetings of the General Conference and the Council and shall make an annual report to both bodies on the work of the Agency and any special reports requested by the General Conference or the Council or which the General Secretary may deem desirable.

5. The General Secretary shall establish the procedures for distributing to all Contracting Parties information received by the Agency from governmental sources and such information from non-governmental sources as may be of interest to the Agency.

6. In the performance of their duties the General Secretary and the staff shall not seek or receive instructions from any Government or from any other authority external to the Agency and shall refrain from any action which might reflect on their position as international officials responsible only to the Agency; subject to their responsibility to the Agency, they shall not disclose any industrial secrets or other confidential information coming to their knowledge by reason of their official duties in the Agency.

7. Each of the Contracting Parties undertakes to respect the exclusively international character of the responsibilities of the General Secretary and the staff and not to seek to influence them in the discharge of their responsibilities.

Article 12. *Control system*

1. For the purpose of verifying compliance with the obligations entered into by the Contracting Parties in accordance with article 1, a control system shall be established which shall be put into effect in accordance with the provisions of articles 13–18 of this Treaty.

2. The control system shall be used in particular for the purpose of verifying:

(*a*) That devices, services and facilities intended for peaceful uses of nuclear energy are not used in the testing or manufacture of nuclear weapons;

(*b*) That none of the activities prohibited in article 1 of this Treaty are carried out in the territory of the Contracting Parties with nuclear materials or weapons introduced from abroad; and

(*c*) That explosions for peaceful purposes are compatible with article 18 of this Treaty.

Article 13. *IAEA safeguards*

Each Contracting Party shall negotiate multilateral or bilateral agreements with the International Atomic Energy Agency for the application of its safeguards to its nuclear activities. Each Contracting Party shall initiate negotiations within a period of 180 days after the date of the deposit of its instrument of ratification of this Treaty. These agreements shall enter into force, for each Party, not later than eighteen months after the date of the initiation of such negotiations except in case of unforeseen circumstances or *force majeure*.

Article 14. *Reports of the Parties*

1. The Contracting Parties shall submit to the Agency and to the International Atomic Energy Agency, for their information, semi-annual reports stating that no activity prohibited under this Treaty has occurred in their respective territories.

2. The Contracting Parties shall simultaneously transmit to the Agency a copy of any report they may submit to the International Atomic Energy Agency which relates to matters that are the subject of this Treaty and to the application of safeguards.

3. The Contracting Parties shall also transmit to the Organization of American States, for its information, any reports that may be of interest to it, in accordance with the obligations established by the Inter-American System.

Article 15. *Special reports requested by the General Secretary*

1. With the authorization of the Council, the General Secretary may request any of the Contracting Parties to provide the Agency with complementary or supplementary information regarding any event or circumstance connected with compliance with this Treaty, explaining his reasons. The Contracting Parties undertake to cooperate promptly and fully with the General Secretary.

2. The General Secretary shall inform the Council and the Contracting Parties forthwith of such requests and of the respective replies.

Article 16. *Special inspections*

1. The International Atomic Energy Agency and the Council established by this Treaty have the power of carrying out special inspections in the following cases:

(*a*) In the case of the International Atomic Energy Agency, in accordance with the agreements referred to in article 13 of this Treaty;

(*b*) In the case of the Council:

(i) When so requested, the reasons for the request being stated, by any Party which suspects that some activity prohibited by this Treaty has been carried out or is about to be carried out, either in the territory of any other Party or in any other place on such latter Party's behalf, the Council shall immediately arrange for such as inspection in accordance with article 10, paragraph 5;

(ii) When requested by any Party which has been suspected of or charged with having violated this Treaty, the Council shall immediately arrange for the special inspection requested in accordance with article 10, paragraph 5.

The above requests will be made to the Council through the General Secretary.

2. The costs and expenses of any special inspection carried out under paragraph 1, sub-paragraph (*b*), sections (i) and (ii) of this article shall be borne by the requesting Party or Parties, except where the Council concludes on the basis of the report on the special inspection that, in view of the circumstances existing in the case, such costs and expenses should be borne by the Agency.

3. The General Conference shall formulate the procedures for the organization and execution of the special inspections carried out in accordance with paragraph 1, sub-paragraph (*b*), sections (i) and (ii) of this article.

4. The Contracting Parties undertake to grant the inspectors carrying out such special inspections full and free access to all places and all information which may be necessary for the performance of their duties and which are directly and intimately connected with the suspicion of violation of this Treaty. If so requested by the authorities of the Contracting Party in whose territory the inspection is carried out, the inspectors designated by the General Conference shall be accompanied by representatives of said authorities, provided that this does not in any way delay or hinder the work of the inspectors.

5. The Council shall immediately transmit to all the Parties, through the General Secretary, a copy of any report resulting from special inspections.

6. Similarly, the Council shall send through the General Secretary to the Secretary-General of the United Nations, for transmission to the United Nations Security Council and General Assembly, and to the Council of the Organization of American States, for its information, a copy of any report resulting from any special inspection carried out in accordance with paragraph 1, sub-paragraph (*b*), sections (i) and (ii) of this article.

7. The Council may decide, or any Contracting Party may request, the convening of a special session of the General Conference for the purpose of considering the reports resulting from any special inspection. In such a case, the General Secretary shall take immediate steps to convene the special session requested.

8. The General Conference, convened in special session under this article, may make recommendations to the Contracting Parties and submit reports to the Secretary-General of the United Nations to be transmitted to the United Nations Security Council and the General Assembly.

Article 17. *Use of nuclear energy for peaceful purposes*

Nothing in the provisions of this Treaty shall prejudice the rights of the Contracting Parties, in conformity with this Treaty, to use nuclear energy for peaceful purposes, in particular for their economic development and social progress.

Article 18. *Explosions for peaceful purposes*

1. The Contracting Parties may carry out explosions of nuclear devices for peaceful purposes – including explosions which involve devices similar to those used in nuclear weapons – or collaborate with third parties for the same purpose, provided that they do so in accordance with the provisions of this article and the other articles of the Treaty, particularly articles 1 and 5.

2. Contracting Parties intending to carry out, or to cooperate in carrying out, such an explosion shall notify the Agency and the International Atomic Energy Agency, as far in advance as the circumstances require, of the date of the explosion and shall at the same time provide the following information:

(*a*) The nature of the nuclear device and the source from which it was obtained;

(*b*) The place and purpose of the planned explosion;

(*c*) The procedures which will be followed in order to comply with paragraph 3 of this article;

(*d*) The expected force of the device; and

(*e*) The fullest possible information on any possible radioactive fall-out that may result from the explosion or explosions, and measures which will be taken to avoid danger to the population, flora, fauna and territories of any other Party or Parties.

3. The General Secretary and the technical personnel designated by the Council and the International Atomic Energy Agency may observe all the preparations, including the explosion of the device, and shall have unrestricted access to any area in the vicinity of the site of the explosion in order to ascertain whether the device and the procedures followed during the explosion are in conformity with the information supplied under paragraph 2 of this article and the other provisions of this Treaty.

4. The Contracting Parties may accept the collaboration of third parties for the purpose set forth in paragraph 1 of the present article, in accordance with paragraphs 2 and 3 thereof.

Article 19. *Relations with other international organizations*

1. The Agency may conclude such agreements with the International Atomic Energy Agency as are authorized by the General Conference and as it considers likely to facilitate the efficient operation of the control system established by this Treaty.

2. The Agency may also enter into relations with any international organization or body, especially any which may be established in the future to supervise disarmament or measures for the control of armaments in any part of the world.

3. The Contracting Parties may, if they see fit, request the advice of the Inter-American Nuclear Energy Commission on all technical matters connected with the application of this Treaty with which the Commission is competent to deal under its Statute.

Article 20. *Measures in the event of violation of the Treaty*

1. The General Conference shall take note of all cases in which, in its opinion, any Contracting Party is not complying fully with its obligations under this Treaty and shall draw the matter to the attention of the Party concerned, making such recommendations as it deems appropriate.

2. If, in its opinion, such non-compliance constitutes a violation of this Treaty which might endanger peace and security, the

General Conference shall report thereon simultaneously to the United Nations Security Council and the General Assembly through the Secretary-General of the United Nations, and to the Council of the Organization of American States. The General Conference shall likewise report to the International Atomic Energy Agency for such purposes as are relevant in accordance with its Statute.

Article 21. *United Nations and Organization of American States*

None of the provisions of this Treaty shall be construed as impairing the rights and obligations of the Parties under the Charter of the United Nations or, in the case of States Members of the Organization of American States, under existing regional treaties.

Article 22. *Privileges and immunities*

1. The Agency shall enjoy in the territory of each of the Contracting Parties such legal capacity and such privileges and immunities as may be necessary for the exercise of its functions and the fulfilment of its purposes.
2. Representatives of the Contracting Parties accredited to the Agency and officials of the Agency shall similarly enjoy such privileges and immunities as are necessary for the performance of their functions.
3. The Agency may conclude agreements with the Contracting Parties with a view to determining the details of the application of paragraphs 1 and 2 of this article.

Article 23. *Notification of other agreements*

Once this Treaty has entered into force, the Secretariat shall be notified immediately of any international agreement concluded by any of the Contracting Parties on matters with which this Treaty is concerned; the Secretariat shall register it and notify the other Contracting Parties.

Article 24. *Settlement of disputes*

Unless the Parties concerned agree on another mode of peaceful settlement, any question or dispute concerning the interpretation or application of this Treaty which is not settled shall be referred to the International Court of Justice with the prior consent of the Parties to the controversy.

Article 25. *Signature*

1. This Treaty shall be open indefinitely for signature by:

(*a*) All the Latin American Republics, and

(*b*) All other sovereign States situated in their entirety south of latitude 35° north in the western hemisphere; and, except as provided in paragraph 2 of this article, all such States which become sovereign, when they have been admitted by the General Conference.

2. The General Conference shall not take any decision regarding the admission of a political entity part or all of whose territory is the subject, prior to the date when this Treaty is opened for signature, of a dispute or claim between an extra-continental country and one or more Latin American States, so long as the dispute has not been settled by peaceful means.

Article 26. *Ratification and deposit*

1. This Treaty shall be subject to ratification by signatory States in accordance with their respective constitutional procedures.

2. This Treaty and the instruments of ratification shall be deposited with the Government of the Mexican United States, which is hereby designated the Depositary Government.

3. The Depositary Government shall send certified copies of this Treaty to the Governments of signatory States and shall notify them of the deposit of each instrument of ratification.

Article 27. *Reservations*

This Treaty shall not be subject to reservations.

Article 28. *Entry into force*

1. Subject to the provisions of paragraph 2 of this article, this Treaty shall enter into force among the States that have ratified it as soon as the following requirements have been met:

(*a*) Deposit of the instruments of ratification of this Treaty with the Depositary Government by the Governments of the States mentioned in article 25 which are in existence on the date when this Treaty is opened for signature and which are not affected by the provisions of article 25, paragraph 2;

(*b*) Signature and ratification of Additional Protocol I annexed to this Treaty by all extra-continental or continental States having *de jure* or *de facto* international responsibility for territories situated in the zone of application of the Treaty;

(*c*) Signature and ratification of the Additional Protocol II annexed to this Treaty by all powers possessing nuclear weapons;

(*d*) Conclusion of bilateral or multilateral agreements on the application of the Safeguards System of the International Atomic Energy Agency in accordance with article 13 of this Treaty.

2. All signatory States shall have the imprescriptible right to waive, wholly or in part, the requirements laid down in the preceding paragraph. They may do so by means of a declaration which shall be annexed to their respective instrument of ratification and which may be formulated at the time of deposit of the instrument or subsequently. For those States which exercise this right, this Treaty shall enter into force upon deposit of the declaration, or as soon as those requirements have been met which have not been expressly waived.

3. As soon as this Treaty has entered into force in accordance with the provisions of paragraph 2 for eleven States, the Depositary Government shall convene a preliminary meeting of those States in order that the Agency may be set up and commence its work.

4. After the entry into force of this Treaty for all countries of the zone, the rise of a new power possessing nuclear weapons shall have the effect of suspending the execution of this Treaty for those countries which have ratified it without waiving requirements of paragraph 1, sub-paragraph (*c*) of this article, and which request such suspension; the Treaty shall remain suspended until the new power, on its own initiative or upon request by the General Conference, ratifies the annexed Additional Protocol II.

Article 29. *Amendments*

1. Any Contracting Party may propose amendments to this Treaty and shall submit its proposals to the Council through the General Secretary, who shall transmit them to all the other Contracting Parties and, in addition, to all other signatories in accordance with article 6. The Council, through the General Secretary, shall immediately following the meeting of signatories convene a special session of the General Conference to examine the proposals made, for the adoption of which a two-thirds majority of the Contracting Parties present and voting shall be required.

2. Amendments adopted shall enter into force as soon as the requirements set forth in article 28 of this Treaty have been complied with.

Article 30. *Duration and denunciation*

1. This Treaty shall be of a permanent nature and shall remain in force indefinitely, but any Party may denounce it by notifying the General Secretary of the Agency if, in the opinion of the denouncing State, there have arisen or may arise circumstances connected with the content of this Treaty or of the annexed Additional Protocols I and II which affect its supreme interests or the peace and security of one or more Contracting Parties.

2. The denunciation shall take effect three months after the delivery to the General Secretary of the Agency of the notification by the Government of the signatory State concerned. The General Secretary shall immediately communicate such notification to the other Contracting Parties and to the Secretary-General of the United Nations for the information of the United Nations Security Council and the General Assembly. He shall also communicate it to the Secretary-General of the Organization of American States.

Article 31. *Authentic texts and registration*

This Treaty, of which the Spanish, Chinese, English, French, Portuguese and Russian texts are equally authentic, shall be registered by the Depositary Government in accordance with article 102 of the United Nations Charter. The Depositary Government shall notify the Secretary-General of the United Nations of the signatures, ratifications and amendments relating to this Treaty and shall communicate them to the Secretary-General of the Organization of American States for its information.

Transitional Article

Denunciation of the declaration referred to in article 28, paragraph 2, shall be subject to the same procedures as the denunciation of this Treaty, except that it will take effect on the date of delivery of the respective notification.

In witness whereof the undersigned Plenipotentiaries, having deposited their full powers, found in good and due form, sign this Treaty on behalf of their respective Governments.

Done at Mexico, Distrito Federal, on the Fourteenth day of February, one thousand nine hundred and sixty-seven.

ADDITIONAL PROTOCOL I

The undersigned Plenipotentiaries, furnished with full powers by their respective Governments,

Convinced that the Treaty for the Prohibition of Nuclear Weapons in Latin America, negotiated and signed in accordance with the recommendations of the General Assembly of the United Nations in Resolution 1911 (XVIII) of 27 November 1963, represents an important step towards ensuring the non-proliferation of nuclear weapons,

Aware that the non-proliferation of nuclear weapons is not an end in itself but, rather, a means of achieving general and complete disarmament at a later stage, and

Desiring to contribute, so far as lies in their power, towards ending the armaments race, especially in the field of nuclear weapons, and towards strengthening a world at peace, based on mutual respect and sovereign equality of States,

Have agreed as follows:

Article 1

To undertake to apply the statute of denuclearization in respect of warlike purposes as defined in articles 1, 3, 5 and 13 of the Treaty for the Prohibition of Nuclear Weapons in Latin America in territories for which, *de jure* or *de facto*, they are internationally responsible and which lie within the limits of the geographical zone established in that Treaty.

Article 2

The duration of this Protocol shall be the same as that of the Treaty for the Prohibition of Nuclear Weapons in Latin America of which this Protocol is an annex, and the provisions regarding ratification and denunciation contained in the Treaty shall be applicable to it.

Article 3

This Protocol shall enter into force, for the States which have ratified it, on the date of the deposit of their respective instruments of ratification.

In witness whereof the undersigned Plenipotentiaries, having deposited their full powers, found in good and due form, sign this Protocol on behalf of their respective Governments.

ADDITIONAL PROTOCOL II

The undersigned Plenipotentiaries, furnished with full powers by their respective Governments,

Convinced that the Treaty for the Prohibition of Nuclear Weapons in Latin America, negotiated and signed in accordance with the recommendations of the General Assembly of the United Nations in Resolution 1911 (XVIII) of 27 November 1963, represents an important step towards ensuring the non-proliferation of nuclear weapons,

Aware that the non-proliferation of nuclear weapons is not an end in itself but, rather, a means of achieving general and complete disarmament at a later stage, and

Desiring to contribute, so far as lies in their power, towards ending the armaments race, especially in the field of nuclear weapons, and towards promoting and strengthening a world at peace, based on mutual respect and sovereign equality of States,

Have agreed as follows:

Article 1

The statute of denuclearization of Latin America in respect of warlike purposes, as defined, delimited and set forth in the Treaty for the Prohibition of Nuclear Weapons in Latin America of which this instrument is an annex, shall be fully respected by the Parties to this Protocol in all its express aims and provisions.

Article 2

The Governments represented by the undersigned Plenipotentiaries undertake, therefore, not to contribute in any way to the performance of acts involving a violation of the obligations of article 1 of the Treaty in the territories to which the Treaty applies in accordance with article 4 thereof.

Article 3

The Governments represented by the undersigned Plenipotentiaries also undertake not to use or threaten to use nuclear weapons against the Contracting Parties of the Treaty for the Prohibition of Nuclear Weapons in Latin America.

Article 4

The duration of this Protocol shall be the same as that of the Treaty for the Prohibition of Nuclear Weapons in Latin America of which this Protocol is an annex, and the definitions of territory and nuclear weapons set forth in articles 3 and 5 of the Treaty shall be applicable to this Protocol, as well as the provisions regarding ratification, reservations, denunciation, authentic texts and registration contained in articles 26, 27, 30 and 31 of the Treaty.

Article 5

This Protocol shall enter into force, for the States which have ratified it, on the date of the deposit of their respective instruments of ratification.

In witness whereof, the undersigned Plenipotentiaries, having deposited their full powers, found to be in good and due form, hereby sign this Additional Protocol on behalf of their respective Governments.

Source: *Treaty Series*, Vol. 634 (United Nations, New York, 1970).

Appendix III

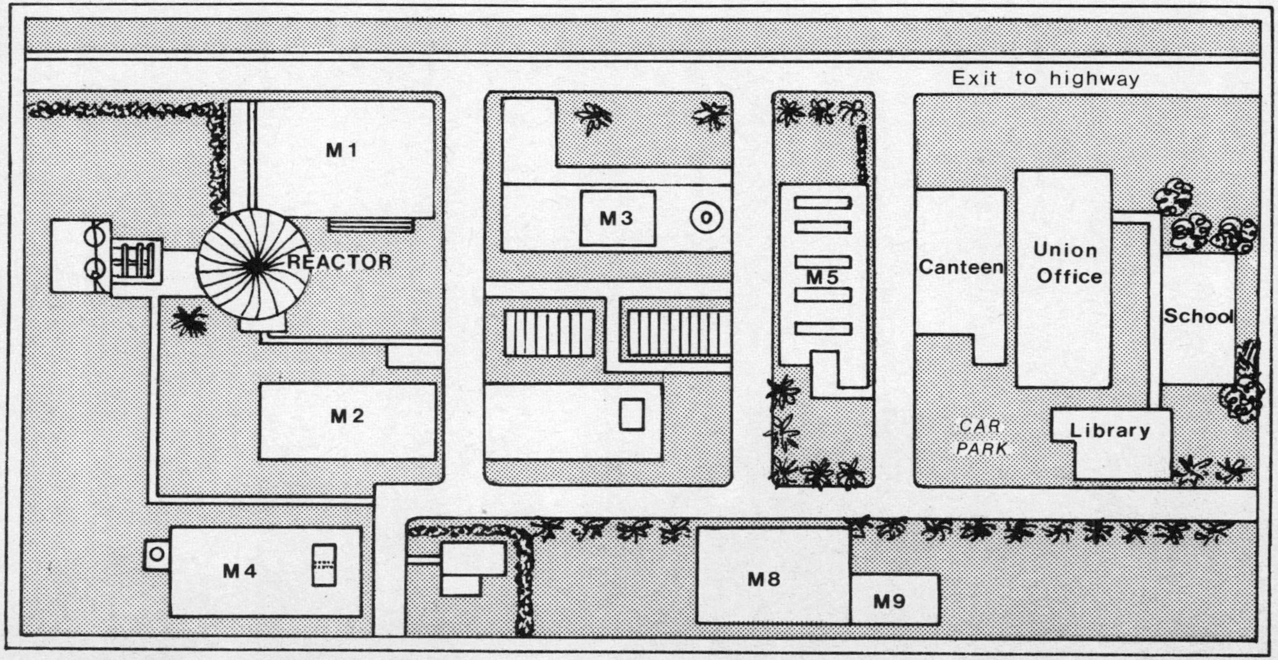

Ground Plan of Dimona

References

Adams, James (1984), *The Unnatural Alliance* (Quartet Books, New York).

Albright, David (1987a), 'Civilian Inventories of Plutonium and Highly Enriched Uranium' in Leventhal, Paul and Yonah Alexander (eds), *Preventing Nuclear Terrorism* (Lexington Books, Lexington, Mass.).

—— (1987b) 'Pakistan's Bomb-making Capacity', *Bulletin of the Atomic Scientists*, Vol. 43, No. 5 (June).

Beit-Hallahmi, Benjamin (1988), *The Israeli Connection* (I.B. Tauris, London).

Beker, Avi (1986), 'A Regional Non-Proliferation Treaty for the Middle East' in Beres (ed.), *Security or Armageddon*.

Beres, Louis René (ed.) (1986), *Security or Armageddon: Israel's Nuclear Strategy* (Lexington Books, Lexington, Mass.).

Bhatia, Shyam (1988), *Nuclear Rivals in the Middle East* (Routledge, London).

Bhutto, Ali (1969), *The Myth of Independence* (Oxford University Press, Karachi).

Boskma, Peter (1974), 'Uranium Enrichment Technologies and the Demand for Enriched Uranium' in Stockholm International Peace Research Institute, *Nuclear Proliferation Problems* (SIPRI, Stockholm).

Brzoska, Michael and Thomas Ohlsen (eds) (1986), *Arms Production and the Third World* (Taylor & Francis, London).

Calvocoressi, Peter (1987), *A Time for Peace* (Hutchinson, London).

Carnegie Endowment for International Peace (1988), *Nuclear Weapons and South Asian Security*, Report of the Carnegie Task Force on Non-Proliferation and South Asian Security (Washington D.C.).

Cervenka, Zdenek and Barbara Rogers (1978), *The Nuclear Axis* (Times Books, New York).

Cohen, Aharon (1970), *Israel and the Arab World* (W.H. Allen, London).

Cohen, Avner (1986), 'Deterrence, Holocaust, and Nuclear Weapons: A Nonparochial Outlook' in Beres (ed.), *Security or Armageddon.*

Davenport, Elaine, Paul Eddy and Peter Gillman (1978), *The Plumbat Affair* (Lippincott, Philadelphia).

Davies, Owen (1982) *The Illegality of Nuclear Warfare* (Lawyers Versus the Bomb, London).

Donnelly, Warren (1986), *Israel's Interest in Nuclear Power: Implications for U.S. Non-proliferation Policy*, Issue Brief (Congressional Research Service, Washington D.C.).

Executive Office of the President, Office of Science and Technology Policy (1980), *Ad Hoc Panel Report on the September 22 Event* (Washington D.C.).

Feldman, Shai (1982), *Israeli Nuclear Deterrence* (Columbia University Press, New York).

Freedman, Lawrence (1975), 'Israel's Nuclear Policy', *Survival*, (June).

Freier, Shalhevet (1987), *Israeli Nuclear Armament*, Statement to U.N. First Committee, 2 November.

de Gaulle, Charles (1970), *Memoires d'Espoir – Le Renouveau 1958–1962* (Librairie Plon, Paris).

Goldblat, Jozef (ed.) (1985), *Non-Proliferation: the Why and the Wherefore* (Taylor & Francis, London).

Gowing, Margaret (1974), *Independence and Deterrence* (Macmillan, London).

Harkavy, Robert (1986), 'The Imperative to Survive' in Beres (ed.), *Security or Armageddon.*

Inbar, Ephraim (1986) 'Israel and Nuclear Weapons since October 1973' in Beres (ed.), *Security or Armageddon.*

Jabber, Fuad (1971), *Israel and Nuclear Weapons* (Chatto & Windus, London).

Kapur, Ashok (1985), 'Pakistan' in Goldblat (ed.), *Non-Proliferation.*

Kats, Gregory H. (1985), 'Egypt' in Goldblat (ed.), *Non-Proliferation.*

Kellen, Konrad (1987), 'The Potential for Nuclear Terrorism: A Discussion' in Leventhal and Alexander (eds), *Preventing Nuclear Terrorism.*

Lovins, Amory B. (1980), 'Nuclear Weapons and Power-reactor Plutonium', *Nature*, 28 February 1980.

Lee, John Marshall (1984), *The Use of Nuclear Weapons* (Hubert

H. Humphrey Institute, University of Minnesota).

Mark, J. Carson, Theodore Taylor, Eugene Eyster, William Maraman and Jacob Wechsler (1987), 'Can Terrorists Build Nuclear Weapons' in Leventhal and Alexander (eds), *Preventing Nuclear Terrorism.*

Milhollin, Gary (1986), *Israel's Nuclear Shadow* (Wisconsin Project on Nuclear Arms Control, Wisconsin).

—— (1987), 'Heavy Water Cheaters', *Foreign Policy*, No. 69 (Winter).

Myrdal, Alva (1976), *The Game of Disarmament* (Pantheon Books, New York).

Nucleonics Week, 'Pakistan and China Signed a Comprehensive Agreement for Co-operation', 25 September.

Office of Technology Assessment, U.S. Congress (1977), *Nuclear Proliferation and Safeguards* (Washington D.C.).

Pajak, Roger F. (1982), *Nuclear Proliferation in the Middle East* (National Defense University Press, Washington D.C.).

Pry, Peter (1984), *Israel's Nuclear Arsenal* (Croom Helm, London).

Rogers, Bernard (1983), Statement to the Dutch Parliament, The Hague.

Rotblat, Joseph (1981), *Nuclear Radiation in Warfare* (Taylor & Francis, London).

Sayigh, Yezid (1983), 'Palestine Military Performance in the 1982 War', *Journal of Palestine Studies*, Vol. XII, No. 4 (Summer).

Segal, David (1987), 'Iran: In a Hurry to Have Their Bomb', *International Herald Tribune*, 14 April; and 'Iran: The Mullahs are Playing Dangerous Games with Nuclear Fire', *International Herald Tribune*, 15 April.

Senate Foreign Relations Committee (1981), *Hearings on the Israeli Air Strike*, 97th Congress, 1st Session, 18, 19 and 25 June (Government Printing Office, Washington D.C.).

Spector, Leonard (1985), *The New Nuclear Nations* (Vintage Books, New York).

—— (1987), *Going Nuclear* (Ballinger, Cambridge, Mass.).

Stockholm International Peace Research Institute (SIPRI) (1979), *World Armaments and Disarmament, SIPRI Yearbook 1979* (Taylor & Francis, London).

United Nations (1982) (1987), *Study on Israeli Nuclear Armament* (New York).

US Department of Defense (1987), *World Commerce in Nuclear Materials* (Washington DC.).

Van Doren, Charles (1987), 'Pakistan and Congress', *Arms Control Today*, Vol. 17, No. 9.

Van Leeuwen, Marianne and Ben Soetendorp (1987) 'Israel', in Muller, Harald (ed.) *A European Non-Proliferation Policy* (Oxford University Press, Oxford).

Washington Office on Africa Educational Fund (1985), *The September 22, 1979 Mystery Flash: Did South Africa Detonate a Nuclear Bomb?* (Washington D.C.).

Weissman, Steve and Herbert Krosney (1981), *The Islamic Bomb* (Times Books, New York).

Willrich, Mason and Theodore B. Taylor (1974), *Nuclear Theft: Risks and Safeguards* (Ballinger, Cambridge, Mass.).

York, Herbert (1973), *Deterrence by Means of Mass Destruction* (Pacem in Terris Conference III, Washington D.C.).

Index